Stability of Nonlinear Systems

ELECTRONIC & ELECTRICAL ENGINEERING RESEARCH STUDIES

CONTROL THEORY AND APPLICATIONS STUDIES SERIES

Series Editor: **Professor M.J.H. Sterling**
Department of Engineering Science,
University of Durham,
Durham, England

1. Stability of Nonlinear Systems
Professor Derek P. Atherton

Stability of Nonlinear Systems

Professor Derek P. Atherton, B. Eng., Ph.D., D. Sc.
School of Engineering and Applied Sciences,
The University of Sussex, England

RESEARCH STUDIES PRESS
A DIVISION OF JOHN WILEY & SONS LTD
Chichester · New York · Brisbane · Toronto

RESEARCH STUDIES PRESS

Editorial Office:
8 Willian Way, Letchworth, Herts SG6 2HG, England

British Library Cataloguing in Publication Data:

Atherton, Derek Percy
Stability of nonlinear systems. - (Electronic and Electrical engineering research studies: control theory and applications studies series; vol. 1).
1. System analysis
2. Nonlinear oscillations
I. Title II. Series
003 QA402 80-40947

ISBN 0 471 27856 4

Printed in the United States of America

CONTENTS

EDITORIAL PREFACE

For many years the field of control systems analysis and design has attracted substantial research effort. The extension of classical single variable techniques to the multivariable case has often necessitated practically unrealistic assumptions concerning plant behaviour, especially in respect of non-linearities. Even techniques developed specifically for multivariable systems analysis can be too restrictive in the type of nonlinear elements which can be considered. Unfortunately from an analysis standpoint, nonlinear plant is all too common and consequently this volume aims to provide an insight into the techniques currently available for the analysis and controller design in non-linear systems.

The material included provides a survey of frequency response type methods which can be used for investigation of the stability of autonomous non-linear feedback systems. The inclusion of worked examples is designed to aid the understanding of what, to the uninitiated, often seems to be abstract mathematics. Topics covered include absolute stability criteria, the describing function method, the Aizerman and Kalman conjectures, together with limit cycles in relay systems. The final chapter deals with multivariable systems and presents some extensions of single variable methods which result in practical analysis and design techniques.

This volume represents the first in a series of research level monographs under the general title of "Control Theory and Applications Studies". The series aims to provide for the dissemination of research results in all aspects of control engineering and related topics, both in theoretical aspects and also in applications likely to be of wide interest.

M.J.H. Sterling

PREFACE

The aim of this book is to give an overview of the techniques available for investigating the stability of an autonomous nonlinear feedback system. It is written so that it can be followed by a reader who has done a first course in automatic control theory and can be used as a text for undergraduate or postgraduate courses which include material on nonlinear systems. Since the book is primarily concerned with the application of the techniques presented no elaborate mathematical proofs of results, particularly in the area of absolute stability, are given. The interested reader can find these additional details by consulting the references cited. The adoption of this approach has enabled the author to keep the book to an acceptable size for a monograph and also makes it of value to the industrial designer who is interested more in the applicability of the various procedures than the mathematical rigour of their derivations.

Chapter 1 discusses nonlinear systems, the definitions of stability and the relationship of instability to the existence of limit cycles. Second order systems, both because of their historical importance and the general insight they provide for the understanding of more complex systems, are considered in Chapter 2. Particular emphasis is placed on the graphical procedures of the phase plane.

The advent of interactive computing and the availability of graphics terminals have enabled the iterative methods of the frequency domain design procedures of classical control theory to be applied more efficiently. For this reason and because of the lack of easily applied time domain methods the remaining chapters concentrate on frequency domain procedures. In Chapter 3 the most useful available absolute stability criteria are presented and their applicability to various problems discussed. Chapter 4 presents the describing function, DF, method for investigating stability and the evaluation of limit cycles. The evaluation of asymmetrical limit cycles and the determination of the stability of a limit cycle are also discussed. Because the DF method is approximate it is important to have procedures for validating the accuracy of any solutions. This problem is discussed in Chapter 5 together with a discussion of the linearization conjectures of Aizerman and Kalman. The latter sections of the chapter discuss various extensions of the DF method including its use for determining combined mode oscillations.

The evaluation of limit cycles in relay control systems by an exact method, originally due to Tsypkin, is discussed in Chapter 6. An extension of the method is given to cover some unusual forms of oscillation with multiple pulses per half period. Finally Chapter 7 considers extensions of the methods of chapters 3, 4 and 6 to the study of multivariable nonlinear systems. Although the use of computer graphics is helpful for the techniques of the early chapters, the complexity of the methods of Chapter 7 is such that without this type of facility the procedures become extremely difficult to apply. Several worked examples are included in the text and some problems are given at the end of each chapter.

I am indebted to several students for both reading and commenting on the material presented and, in particular, to my colleague Dr. Balasubramanian for reading the entire manuscript and providing helpful comments and advice. I also wish to thank Georgi May and Edith LaBillois for their rough typing of the manuscript, Susan Perrott for typing the final version and Stan Bowser for drawing the figures.

Fredricton, July 1980

CHAPTER 1
Introduction

1.1 REPRESENTATIONS OF DYNAMICAL SYSTEMS

The study of a given physical system starts with the construction of a mathematical model. The mathematical model which describes the given system in the best manner depends upon physical factors as well as the objectives of the mathematical modelling. In control systems, the physical factors include ranges of the variables, the system environment, assumptions regarding the various components of the system and how they are interconnected. To simplify analysis of the resulting model we often assume that the physical relationships are linear. Since nonlinear effects always come into play in real systems, the linear model will normally only be satisfactory within some range of the system variables. The choice of the technique adopted for modelling the system will depend upon the subsequent use to which the model is put. Such modelling techniques include mathematical, qualitative, analytical or computational. In turn, the choice of the model will influence the approach used, since the simpler the model the easier, for example, it will be to use an analytical approach. The goodness of a mathematical model on the other hand is judged by the closeness with which its predicted behaviour describes the behaviour of the physical system it represents.

The stability of a linear model is easily determined by one of several well known techniques such as the Routh-Hurwitz criterion, root locus method, or the Nyquist criterion. The conditions which these approaches give are both necessary and sufficient for stability of the linear system. They are also valid in a global sense since, unlike a nonlinear system, the stability of a linear system is not affected by initial conditions or external inputs. The two most common mathematical models used for linear systems are the state space and transfer function representations. In the former the single input-single output system is described by the state equations

$$\begin{aligned} \dot{x} &= Ax + bu \\ y &= c^T x + du \end{aligned} \qquad (1.1)$$

where the state vector x is nx1, A is an nxn matrix and b and c^T are nx1 vectors. The transfer function, which is the Laplace transform of the system impulse response, describes the system on an input-output basis according to the relationship,

$$Y(s) = G(s)U(s) \qquad (1.2)$$

where $Y(s)=\mathcal{L}\{y(t)\}$ and $U(s)=\mathcal{L}\{u(t)\}$. It is easily shown from eq. (1.1) that

$$G(s) = c^T(sI - A)^{-1}b \qquad (1.3)$$

and provided the system is controllable and observable the poles of G(s) are the eigenvalues of A. When the eigenvalues of A all lie in the left hand side of the s plane A is Hurwitz.

In modelling control systems one often develops the complete system model from models determined for the individual components. When this is the case it may be possible to identify the model in terms of a static nonlinear operation and linear dynamics. For this reason and because there is no general analytical approach available for the solution of nonlinear systems, the specific nonlinear model shown in the block diagram of Fig. 1.1 is of prime concern. Although some of the analysis techniques to be presented in this monograph can be used for more complex situations most of the exact techniques for stability investigation are directed to the study of this particular system. Because the majority of techniques for investigating the stability of more sophisticated configurations are approximate, comparison of the approximate methods with rigorous mathematical results for the system of Fig. 1.1 provides an opportunity for understanding the basic mechanisms of nonlinear behaviour and the limitations of approximate methods of analysis. Further it should be pointed out that the introduction of the single nonlinear element does severely complicate the analysis of the system. No exact results giving necessary and sufficient conditions for the stability of the general system exist.

The primary aim of this book is to present methods for the investigation of the stability of the autonomous system, that is with r(t)=0, shown in Fig. 1.1. In chapter 7, however, we expand our horizons somewhat by considering a multivariable version of this block diagram. The emphasis throughout is on the application of the methods to the solution of problems on the stability or occurrence of oscillations in nonlinear systems found in many areas of Engineering and Science.

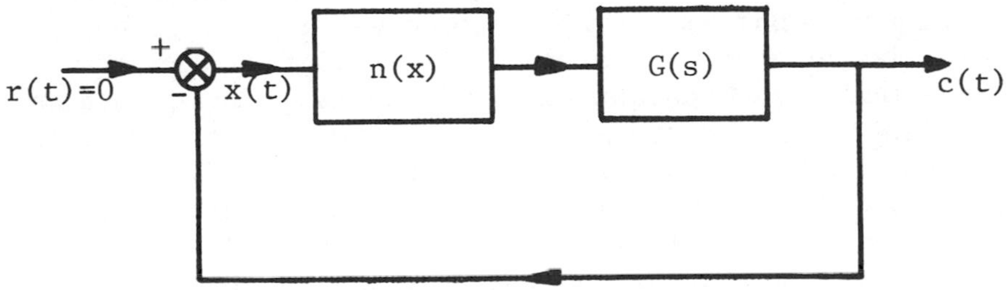

Figure 1.1 The basic system

Despite the possible prevalence of state space techniques during the last two decades frequency response methods have much appeal to industrial designers not least because component models are often available in this domain. The advent of computer graphics facilities within the last few years, which enable frequency response diagrams to be drawn automatically, has added further to the acceptability of this approach. In addition, for nonlinear systems there are often no alternatives to the frequency domain approach apart from direct simulation. Although it is not our purpose in this book to discuss design methods, frequency response techniques are readily usable in the design of nonlinear systems.

1.2 NONLINEAR SYSTEMS

The major difficulty in the consideration of nonlinear systems is that as superposition does not apply, knowing a specific solution to a nonlinear problem may reveal little about the general behaviour of the nonlinear system. On the other hand, for a linear system if, for example, the unit step response is known then in principle the response to any other input can be found. Also all initial condition responses return to the same rest or equilibrium point.

There are many forms of behaviour and phenomena which are distinct to nonlinear systems and further a specific phenomenon may only be exhibited in a particular system provided certain inputs or initial conditions are used. The study of forced nonlinear systems, that is systems with external inputs, which have many interesting aspects, is outside the considerations of this monograph. Here we are concerned with the autonomous system of Fig. 1.1 which in some cases may have constant or bias inputs. The questions we wish to answer are: (i) what form will the system response take for all allowable initial conditions? (ii) are there one or more equilibrium points? (iii) are they stable? (iv) how are they reached? and (v) if they are not stable do the system variables become unbounded, as is the case for a linear system with poles in the right hand side of the s plane, or does an oscillation exist? Before we can answer these questions analytically we must define the terms used.

Consider the autonomous nonlinear system with state space equation

$$\dot{x} = f(x) \tag{1.4}$$

where x is an n vector and assume that it has an equilibirum state or null solution at the origin of the state space, that is $\dot{x}=0$ at $x=0$. This is quite general since if an equilibrium point existed elsewhere in the state space it could be moved to the origin by a simple transformation of coordinates. The definitions of stability, due to Lyapunov [1], are as follows.

Definition 1

The equilibrium state 0, or the equilibrium solution $x(t)=0$, is called stable if for any given positive ε, there exists a positive δ such that

$$||x(t_0)|| < \delta$$

implies

$$||x(t)|| < \varepsilon$$

for all $t > t_0$, where $||\ ||$ denotes the Euclidean norm.

It should be noted that a system exhibiting an oscillation or solutions for which $||x(t)||$ increases temporarily will be considered stable by this definition, when ε and δ are appropriately chosen, but it rules out solutions which grow without bound. Typical situations are illustrated in Fig. 1.2. For a nonlinear system stability may only exist within some domain $||x|| < R$ of the state space.

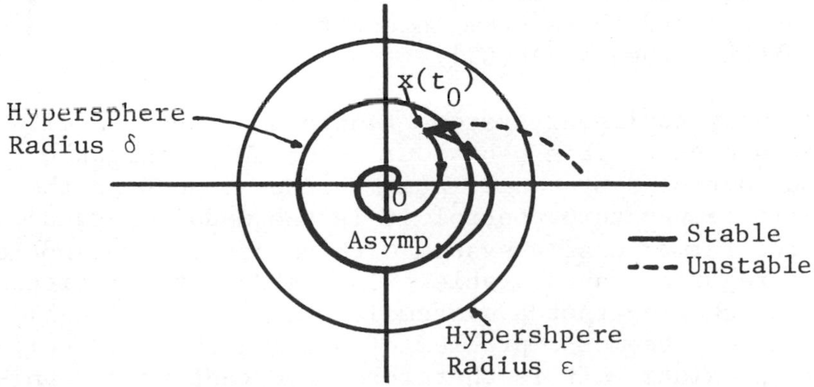

Figure 1.2 Illustration of stability definitions

Definition 2

The equilibrium state 0 is called asymptotically stable, if it is stable and $x(t) \to 0$ as $t \to \infty$.

Definition 3

The equilibrium state 0 is called asymptotically stable in the large, or globally asymptotically stable if it is asymptotically stable and the domain R covers the whole state space.

Many investigations of stability are concerned with the behaviour of forced systems and in this case a more appropriate definition of stability is one which considers bounds on the magnitude of the system variables rather than the properties of an equilibrium state [2].

Definition 4

A signal $x(t) \in \{L_2\}$ if

$$\int_0^\infty x^2(t)\, dt < \infty$$

Definition 5

A system is bounded input-bounded output (BIBO) stable if the input, $r(t) \in \{L_2\}$ and output, $c(t) \in \{L_2\}$.

Remark

For the linear system of eq. (1.1) Definitions 1 and 5 are equivalent. Also for eq. (1.1) if $u(t) \in \{L_2\}$, $y(t) \in \{L_2\}$ and the system is controllable and observable then $x(t) \in \{L_2\}$.

In many nonlinear systems finite amplitude oscillations occur. This behaviour is often undesirable in a control system and in some others, such as the on-off temperature control it is the modus operandi. In the former case a system with an oscillation would be categorized as unstable. In most practical situations, showing that a system is stable, as we shall see later becomes a problem of proving that no oscillation exists. It is therefore important to recognize that Definitions 1 and 5, assuming for the latter case that the average value of c(t) is zero, class an oscillation as stable.

When an oscillation is predicted it may be stable or unstable.

Definition 6

An oscillatory motion x(t) is stable if for a small perturbation the new motion x*(t) is such that $|x^*(t) - x(t)| < \delta$ for all t where δ is small.

For more detailed mathematical considerations the reader is referred to reference [3].

1.3 OUTLINE OF THE BOOK

Chapter 2 is devoted to the study of second order systems. Although the techniques presented cannot in general be extended to higher order systems, they do provide a basic understanding of many of the properties of nonlinear systems. It is also true that there are many systems for which a second order representation proves an adequate mathematical model. The first part of the chapter considers analytical methods which give some valuable insight and are also important from an historical viewpoint. It is interesting to reflect on the effort which has gone into developing some of these methods to obtain a specific initial condition solution which can now be obtained in a matter of seconds on a modern analog or digital computer. The phase plane approach is next considered since it provides an excellent understanding of how solutions can be appreciably different when initial conditions are changed in a nonlinear system. The techniques of chapter 2 are still valuable with modern computational techniques as they enable one to obtain quickly the various regions in which the response can appear significantly different.

Chapter 3 discusses absolute stability methods beginning with Liapunov's method but later concentrating on the various frequency domain criteria. The aim of the chapter is to collect together the known frequency domain results and present them in such a manner that they can be easily applied by the practicing engineer. For this reason and in order to keep the text to reasonable size no proofs, which can in most cases be obtained either using Liapunov's method or by functional analysis, are given.

Chapter 4 discusses the describing function, DF, method and shows how it can be used for investigating the stability of systems which operate either symmetrically or asymmetrically. The incremental describing function method is presented as a procedure for testing whether any predicted limit cycles are stable or unstable.

The following chapter, Chapter 5, contains additional information on the DF method and a discussion of the Aizerman and Kalman conjectures. Various approaches for assessing the accuracy of the describing function method are given. These results clearly show that the describing function method is inappropriate for the analysis of those systems which violate the Aizerman conjecture. Later in the chapter we discuss applications of the DF method to systems with frequency dependent nonlinearities and to systems with more than one nonlinear element in the feedback loop. Also discussed briefly is the problem of combined oscillations. The difficulty encountered using the DF in these more complicated situations is the problem of justifying its validity, since in principle the procedure can be used with any number of nonlinear and linear elements in the system. This problem is not considered in this text although it is clear that the distortion criteria introduced earlier in the chapter can be applied to more complicated systems.

Chapter 6 considers the exact determination of limit cycles in relay systems using Tsypkin's method. An advantage of this approach is that if a limit cycle of the assumed form exists then the method obtains its exact parameters. In this sense it provides a useful reference for comparison with the DF approach. It is also shown that in systems with complicated transfer functions unusual types of periodic oscillation can exist. In these cases a difficulty with the Tsypkin approach is the determination of the number of relay switchings in the periodic waveform.

The final chapter looks at the application of the methods considered previously to multivariable systems. The multivariable configuration considered is the multivariable version of the block diagram of Fig. 1.1. Although most of the methods described in the earlier chapters have been extended to the multivariable situation, further complications arise. In the case of absolute stability methods, they tend to become even more conservative whilst the describing function method requires additional assumptions regarding the oscillation frequencies in the various loops. Similar comments regarding the possible types of limit cycle which can exist apply to the applicability of the Tsypkin method. It is clear that to use any of these approaches one must have a good understanding of the probable system behaviour. In this context there is no substitute for experience and therefore the need to fully understand the methods presented for simpler systems in the early chapters.

A reasonable number of references has been included in each chapter; although a more detailed bibliography can be found in reference [4]. Several worked examples are included in the text and additional problems are given at the end of the chapters.

REFERENCES

1. Liapunov, A.M.: "Stability in nonlinear control systems", Princeton University Press, Princeton, N.J., 1961.

2. Vidyasagar, M.: "Nonlinear systems analysis", Prentice-Hall, N.J., 1978.

3. Struble, R.A.: "Nonlinear differential equations" McGraw-Hill, N.Y., 1962.

4. Atherton D.P. and Dorrah, H.T.: "A survey on non-linear oscillations", Int. J. Control, Vol. 31, pp 1041-1105, 1980.

CHAPTER 2

Second Order Systems

2.1 INTRODUCTION

It is appropriate to begin a study of the stability of nonlinear systems by considering a special category, namely those which can be described by the first order coupled differential equations

$$\begin{aligned} \dot{z}_1 &= f_1(z_1,z_2) \\ \dot{z}_2 &= f_2(z_1,z_2) \; . \end{aligned} \tag{2.1}$$

Although the material presented in this chapter is primarily restricted to second order systems, its importance should not be underestimated. Since many systems of interest in science and engineering can be approximated by eqs. (2.1) there has been a tremendous volume of research devoted to their study. The literature includes original contributions derived from investigations, for example, by Poincaré [1] in celestial mechanics, Rayleigh [2] in vibrations, van der Pol [3] on nonlinear oscillators and Volterra [4] in biology. The fact that graphical methods can be used to study second order systems provides an excellent approach for illustrating concepts which carry over to higher order systems. Further from the control systems viewpoint the method can be used for more general systems than Fig. 1.1 as in principle no restriction is placed on the number of nonlinear elements. Eqs. (2.1) are a nonlinear version of the autonomous state space equation

$$\dot{z} = A z \tag{2.2}$$

where $z=(z_1,z_2)^T$ and

$$A = \begin{pmatrix} a & b \\ c & d \end{pmatrix} . \tag{2.3}$$

It is well known [5] that this state equation can be transformed by a suitable choice of state variables to

$$\dot{x} = A_c x \tag{2.4}$$

where $x=(x_1,x_2)^T$ and

$$A_c = \begin{pmatrix} 0 & 1 \\ -a_0 & -a_1 \end{pmatrix} . \tag{2.5}$$

Here the state variables, since x_2 is the derivative of x_1, are known as phase variables and the description represents the second order linear differential equation

$$\ddot{x} + a_1\dot{x} + a_0 x = 0 \tag{2.6}$$

where x_1 has been replaced by x. The corresponding second order nonlinear differential equation can be written

$$\ddot{x} + f(x,\dot{x}) = 0 \tag{2.7}$$

which in phase variable form is

$$\dot{x}_1 = x_2 \tag{2.8}$$

$$\dot{x}_2 = -f(x_1,x_2) \tag{2.9}$$

and is a particular form of eqs. (2.1). Since the equation representing a single loop system of the type shown in Fig. 1.1 can be put in the form of eqs. (2.8) we will mainly be concerned with this representation. Solutions of eqs. (2.8) for various initial conditions are often shown as loci, known as trajectories, on a phase plane. The phase plane has x_2 as ordinate, x_1 as abscissa and time is a parameter on a trajectory. Arrows on a trajectory denote the direction of increasing time. A closed trajectory corresponds to a periodic oscillation or limit cycle.

In the next three sections we discuss briefly some analytical methods for investigating solutions, primarily periodic, of eq. (2.7). The theory of the second order linear oscillator is well known, so that an obvious starting point for the study of nonlinear oscillatory behaviour is to write eq. (2.7) in the form

$$\ddot{x} + \omega_o^2 x = \mu g(x,\dot{x}) \tag{2.9}$$

with μ a small parameter. Several asymptotic methods, that is approximate solutions to this equation which result in the known exact solution in the limit as $\mu \to 0$, have been developed. The two most well known are the perturbation method [1, 6] and the method of slowly varying amplitude and phase [7]. The applicability of these classical methods to control systems problems is limited. On the other hand the phase plane approach discussed in the later sections has significant practical utility, primarily for qualitative rather than quantitative investigation, due to the difficulty of obtaining accurate graphical solutions.

2.2 THE PERTURBATION METHOD

In this method a solution of the form

$$x(t) = \sum_{j=0}^{\infty} \mu^j x_j(t) \tag{2.10}$$

is assumed and substituted in eq. (2.9). Collating terms with equal powers of μ yields a set of differential equations which can be successively solved for

$x_0(t)$, $x_1(t)$ etc. A difficulty of the method is the appearance of terms in the solutions for $x_j(t)$, known as secular terms, which are unbounded. The presence of these terms prevents the determination of the periodic behaviour of the system. Their occurrence, however, is not surprising since if we consider the linear equation

$$\ddot{x} + x = -\mu x \tag{2.11}$$

which has the exact solution

$$x = a \cos(1 + \mu)^{1/2} t \tag{2.12}$$

for the initial conditions $x(0)=a$ and $\dot{x}(0)=0$; then expand the solution about $\mu=0$, one obtains

$$x = a \cos t - (1/2)\mu a t \sin t + \ldots \tag{2.13}$$

This is precisely the solution obtained using the perturbation method with the second term, due to the presence of t, being unbounded.

The basic deficiency of this procedure, the Poisson method, is that it does not make provision for the change in the frequency of oscillation with amplitude, as occurs in practice. To overcome this problem Poincaré suggested replacing the independent variable t by τ/ω, where ω is the unknown frequency of oscillation, and developing the required solution $x(\tau)$ as a power series with respect to the small parameter μ, that is

$$x(\tau) = \sum_{j=0}^{\infty} \mu^j x_j(\tau) \tag{2.14}$$

where the solutions $x_j(\tau)$ will be of period 2π. The unknown frequency ω is also expressed as a power series in μ, that is

$$\omega = \omega_o + \sum_{j=1}^{\infty} \mu^j \omega_j \tag{2.15}$$

Rewriting eq. (2.9) in terms of the derivatives of x with respect to τ, where for convenience we shall continue to use the dot notation for a derivative with respect to τ, gives

$$\omega^2 \ddot{x} + \omega_0^2 x = \mu g(x, \omega\dot{x}) \tag{2.16}$$

which is the equation into which substitutions for $x(\tau)$ and ω are made from eqs. (2.14) and (2.15) respectively. As before by collecting terms in equal powers of μ one can solve successively for $x_0(\tau)$, $x_1(\tau)$ etc.

To illustrate the method we consider Rayleigh's equation

$$\ddot{x} + x = \mu\{\dot{x} - (\dot{x}^3/3)\}. \tag{2.17}$$

Using the Poisson method we obtain

$$(\ddot{x}_0 + \mu\ddot{x}_1 + \mu^2\ddot{x}_2 + \ldots) + (x_0 + \mu x_1 + \mu^2 x_2 + \ldots)$$
$$= \mu\{\dot{x}_0 + \mu\dot{x}_1 + \ldots -(1/3)(\dot{x}_0 + \mu\dot{x}_1 + \ldots)^3\}.$$

Collecting terms in equal powers of μ leads to

$$\ddot{x}_0 + x_0 = 0 \tag{2.18}$$

$$\ddot{x}_1 + x_1 = \dot{x}_0 - (\dot{x}_0^3/3) \tag{2.19}$$

$$\ddot{x}_2 + x_2 = \dot{x}_1 - \dot{x}_1\dot{x}_0^2 \tag{2.20}$$

and so on.

For the initial conditions $x(0)=a$ and $\dot{x}(0)=0$ the solution of eq. (2.18) is $x_0 = a \cos t$. Substituting this value in eq. (2.19) and solving for x_1 gives

$$x_1 = \{a(a^2 - 4)/8\}(\sin t - t \cos t) - (a^3/32) + (a^3/96) \sin 3t \tag{2.21}$$

which is unsatisfactory because of the secular term $t \cos t$. To remove the secular term we use eqs. (2.14) and (2.15) in eq. (2.17) written in the form of eq. (2.16) to give

$$(1 + \mu\omega_1 + \mu^2\omega_2 + \ldots)^2(\ddot{x}_0 + \mu\ddot{x}_1 + \mu^2\ddot{x}_2 + \ldots)$$
$$+ (x_0 + \mu x_1 + \mu^2 x_2 + \ldots) = \mu\{(1 + \mu\omega_1 + \ldots)$$
$$(\dot{x}_0 + \mu\dot{x}_1 + \ldots) - (1 + \mu\omega_1 + \ldots)^3$$
$$(\dot{x}_0 + \mu\dot{x}_1 + \ldots)^3/3\}$$

since $\omega_o=1$. Collecting terms in equal powers of μ gives

$$\ddot{x}_0 + x_0 = 0 \tag{2.22}$$

$$\ddot{x}_1 + x_1 + 2\omega_1\ddot{x}_0 = \dot{x}_0 - (\dot{x}_0^3/3) \tag{2.23}$$

$$\ddot{x}_2 + x_2 + (2\omega_2+\omega_1^2)\ddot{x}_0 + 2\omega_1\ddot{x}_1 = \dot{x}_1 + \omega_1\dot{x}_0 - \omega_1\dot{x}_0^3 - \dot{x}_0^2\dot{x}_1 \tag{2.24}$$

and so on.

Solution of eq. (2.22) with the initial conditions $x_0(0)=a$ and $\dot{x}_0(0)=0$ gives

$$x_0(\tau) = a \cos \tau . \tag{2.25}$$

Substituting this solution in eq. (2.23) yields

$$\ddot{x}_1 + x_1 = 2\omega_1 a \cos \tau - a \sin \tau + (a^3/3) \sin^3 \tau$$

which can be written

$$\ddot{x}_1 + x_1 = 2\omega_1 a \cos \tau - a\{1 - (a^2/4)\} \sin \tau - (a^3/12) \sin 3\tau .$$

To exclude secular terms the coefficients of cos τ and sin τ must be zero which yields $\omega_1=0$ and $a^2=4$. The solution to the resulting equation

$$\ddot{x}_1 + x_1 = -(2/3) \sin 3\tau$$

is

$$x_1(\tau) = A \cos \tau + B \sin \tau + (1/12) \sin 3\tau$$

Using the initial conditions $x_1(0)=0$ and $\dot{x}_1(0)=0$ gives

$$x_1(\tau) = -(1/4)\sin\tau + (1/12)\sin 3\tau \qquad (2.26)$$

and the solution to order μ is

$$x(t) = 2\cos\omega t - (1/4)\sin\omega t + (1/12)\sin 3\omega t$$

with $\omega=1$, since the first frequency correction factor ω_1 is zero. A better approximation, to order μ^2, is obtained by solving eq. (2.24) with the expressions for $x_0(\tau)$ and $x_1(\tau)$ from eqs. (2.25) and (2.26) substituted.

2.3 AVERAGING METHODS

Averaging methods, which were introduced by van der Pol [3] and Krylov and Bogolyubov [7], approximate the nonlinearity in the second order system by a quasilinear model equivalent in the steady state to the DF method to be discussed in chapter 4. Van der Pol assumed that for small μ the solution to eq. (2.9) could be approximated by

$$x(t) = c(t)\cos\omega_o t + b(t)\sin\omega_o t \qquad (2.27)$$

where $c(t)$ and $b(t)$ are slowly varying functions of time. The equivalent approach of Krylov and Bogolyubov, known as the method of slowly varying amplitude and phase, was to assume a solution of the form

$$x(t) = a(t)\cos\{\omega_o t + \phi(t)\}. \qquad (2.28)$$

Differentiating this equation gives

$$\dot{x}(t) = \dot{a}(t)\cos\{\omega_o t + \phi(t)\} - a(t)\{\omega_o + \dot{\phi}(t)\} \sin\{\omega_o t + \phi(t)\} \qquad (2.29)$$

which for small variations in $a(t)$ and $\phi(t)$ we assume to be of the form

$$\dot{x}(t) = -\omega_o a(t)\sin\{\omega_o t + \phi(t)\}. \qquad (2.30)$$

Using this result in eq. (2.29) gives

$$\dot{a}(t) \cos\{\omega_0 t + \phi(t)\} - a(t)\dot{\phi}(t) \sin\{\omega_0 t + \phi(t)\} = 0. \tag{2.31}$$

Differentiating eq. (2.30) to find $\ddot{x}(t)$ yields

$$\ddot{x}(t) = -\dot{a}(t)\omega_0 \sin\{\omega_0 t + \phi(t)\} - \omega_0 a(t)\{\omega_0 + \dot{\phi}(t)\} \cos\{\omega_0 t + \phi(t)\}$$

which can be written

$$\ddot{x} + \omega_0^2 x = -\dot{a}(t)\omega_0 \sin\{\omega_0 t + \phi(t)\} - \omega_0 a(t)\dot{\phi}(t) \cos\{\omega_0 t + \phi(t)\}.$$

Substituting for $\ddot{x} + \omega_0^2 x$ in the system differential eq. (2.9), replacing $\omega_0 t + \phi(t)$ by θ and omitting t from $\dot{a}(t)$ and $\dot{\phi}(t)$ yields

$$-\dot{a}\omega_0 \sin\theta - \omega_0 a\dot{\phi} \cos\theta = \mu g(a\cos\theta, -a\omega_0 \sin\theta). \tag{2.32}$$

Writing eq. (2.31) in a similar manner gives

$$\dot{a}\cos\theta - a\dot{\phi}\sin\theta = 0. \tag{2.33}$$

Solving these two equations for $\dot{a}$ and $\dot{\phi}$ and then obtaining approximations for the solutions by averaging over the period 2π of θ gives

$$\dot{a} \approx -(\mu/2\pi\omega_0) \int_0^{2\pi} g(a\cos\theta, -a\omega_0 \sin\theta) \sin\theta \, d\theta \tag{2.34}$$

and

$$\dot{\phi} \simeq -(\mu/2\pi a\omega_0) \int_0^{2\pi} g(a\cos\theta, -a\omega_0 \sin\theta) \cos\theta \, d\theta. \tag{2.35}$$

For a specific nonlinearity these integrals can be evaluated and the results integrated with respect to time to obtain expressions for $a(t)$ and $\phi(t)$. The solution is then given by eq. (2.28) with these values substituted.

$$\hat{x} = \sum_{n=0}^{m} c_n \phi_n(t) \tag{2.54}$$

where the $\phi_n(t)$ are appropriate linearly independent functions. When the functions $\phi_n(t)$ are orthogonal over some interval (a,b) and the differential eq. (2.53) is linear, it can be shown that the integral of the squared value of the residual, $\varepsilon(t)$ where

$$\varepsilon(t) = f(D,\hat{x},t)$$

over the interval (a,b) will be a minimum if

$$\int_a^b \varepsilon(t)\, \phi_n(t)\, dt = 0 \quad \text{for all n.} \tag{2.55}$$

To obtain an approximate solution for a nonlinear differential equation the method again uses eq. (2.55). When periodic behaviour is to be investigated an appropriate choice for $c_n\phi_n(t)$ is $a_n \cos n\omega t + b_n \sin n\omega t$. Further, for this situation the procedure gives exactly the same result as the harmonic balance method computed to the same value of n. In the harmonic balance method a solution of the form $\hat{x}_1 = a_1 \cos \omega t + b_1 \sin \omega t$ is tried in eq. (2.53) and the coefficients a_1 and b_1 are evaluated by equating the $\cos \omega t$ and $\sin \omega t$ terms in the equation. For better accuracy the process is continued by including the next appropriate higher harmonic, say the third, in $\hat{x}$, that is

$$\hat{x} = a_1 \cos \omega t + b_1 \sin \omega t + a_3 \cos 3\omega t + b_3 \sin 3\omega t.$$

Substituting $\hat{x}$ in eq. (2.53) and equating the coefficients of the $\sin \omega t$, $\cos \omega t$, $\sin 3\omega t$ and $\cos 3\omega t$ terms gives the values of a_1, b_1, a_3, and b_3. Here the values of a_1 and b_1 will be different from those found using $\hat{x}_1$ as the approximate solution. This technique like the Ritz-Galerkin approach can in principle be taken to any number of harmonics but even for relatively simple differential equations, such as eq. (2.17), the calculations become quite involved.

Limit cycle solutions, especially for systems such as eq. (2.9) with small μ, often have almost circular orbits in the phase plane so that expressing the equations of motion in polar coordinates may offer some advantages. For such cases a method presented by Luus [11], who assumes a constant radius and a phase which varies linearly with time over one cycle, can be useful for estimating the average radius of a limit cycle.

2.5 PHASE PLANE TOPOLOGY

To investigate solutions to eqs. (2.1) we note that

$$\frac{dz_2}{dz_1} = \frac{\dot{z}_2}{\dot{z}_1} = \frac{f_2(z_1,z_2)}{f_1(z_1,z_2)} . \tag{2.56}$$

This indicates that the slope of a trajectory passing through any point, say (z_1',z_2'), in the z_1-z_2 state plane is given by $f_2(z_1', z_2')/f_1(z_1',z_2')$. Further the slope will be undefined at this point if $f_1(z_1',z_2')= f_2(z_1',z_2')=0$. Such a point is known as a singular or equilibrium point. The singular points of a second order system are classified according to the behaviour of trajectories in their neighbourhood. For the linear system of eq. (2.2) this depends upon the eigenvalues of A and for the nonlinear system of eq. (2.1) on the eigenvalues of the equation linearized about the singular point.

The slope of a trajectory at any point in the phase plane for the phase variable representation of eq. (2.8) is

$$\frac{dx_2}{dx_1} = \frac{\dot{x}_2}{\dot{x}_1} = \frac{-f(x_1,x_2)}{x_2} \tag{2.57}$$

from which, in particular, we note that the slope is zero for points where $f(x_1,x_2)=0$ and infinite where $x_2=0$. Also in the upper half of the phase plane, that is for $x_2>0$, trajectories will be directed to the right as time increases and to the left in the lower half of the phase plane. When $f(x_1,x_2)= -f(-x_1,-x_2)$, that is the nonlinear function is odd in x_1 and x_2, which is the case for the linear system, the slope of a trajectory is the same, but with direction reversed, at points (x_1,x_2) and $(-x_1,-x_2)$.

2.5.1 Singular points

The linear eq. (2.6) has only one singular point at the origin since $x_1=x_2=0$ is the only solution of $x_2=0$, $-a_0x_1-a_1x_2=0$. If the equation has a constant input U, that is

$$\ddot{x} + a_1\dot{x} + a_0x = U \tag{2.58}$$

then the singular point is moved from the origin to $(U/a_0,0)$. The general solution to eq. (2.6) is

$$x_1 = A'e^{\lambda_1 t} + B'e^{\lambda_2 t} \tag{2.59}$$

and

$$x_2 = A'\lambda_1 e^{\lambda_1 t} + B'\lambda_2 e^{\lambda_2 t} \tag{2.60}$$

where λ_1 and λ_2 are the eigenvalues (assumed unequal) of A_c, and A' and B' are constants determined by the initial conditions. The equation of the corresponding phase plane trajectory can in principle be obtained by eliminating t from these equations. The shape of the trajectory near the origin, 0, is dependent on the locations of the eigenvalues λ_1 and λ_2 in the complex plane, which are themselves dependent on the values of a_0 and a_1. If $a_0 \neq 0$ six cases exist which are summarized in Table 2.1. Typical phase portraits corresponding to these situations are shown in Fig. 2.1. A few points should be noted: -

(a) The trajectories spiral into or from a focus and the smaller $|\mathrm{Re}\,\lambda|$, that is the less the damping when $\mathrm{Re}\,\lambda < 0$, the more oscillatory the behaviour.

(b) The two lines $x_2=\lambda_1x_1$ and $x_2=\lambda_2x_1$, which are the eigenvector directions of the A_c matrix, are possible trajectories to or from the node. Trajectories cannot cross these lines. For the stable node all trajectories, apart from any starting on $x_2=\lambda_2x_1$ tend to the node along $x_2=\lambda_1x_1$, the slow eigenvector, assuming $|\lambda_1|<|\lambda_2|$. When $\lambda_1=\lambda_2$ the two eigenvectors overlap.

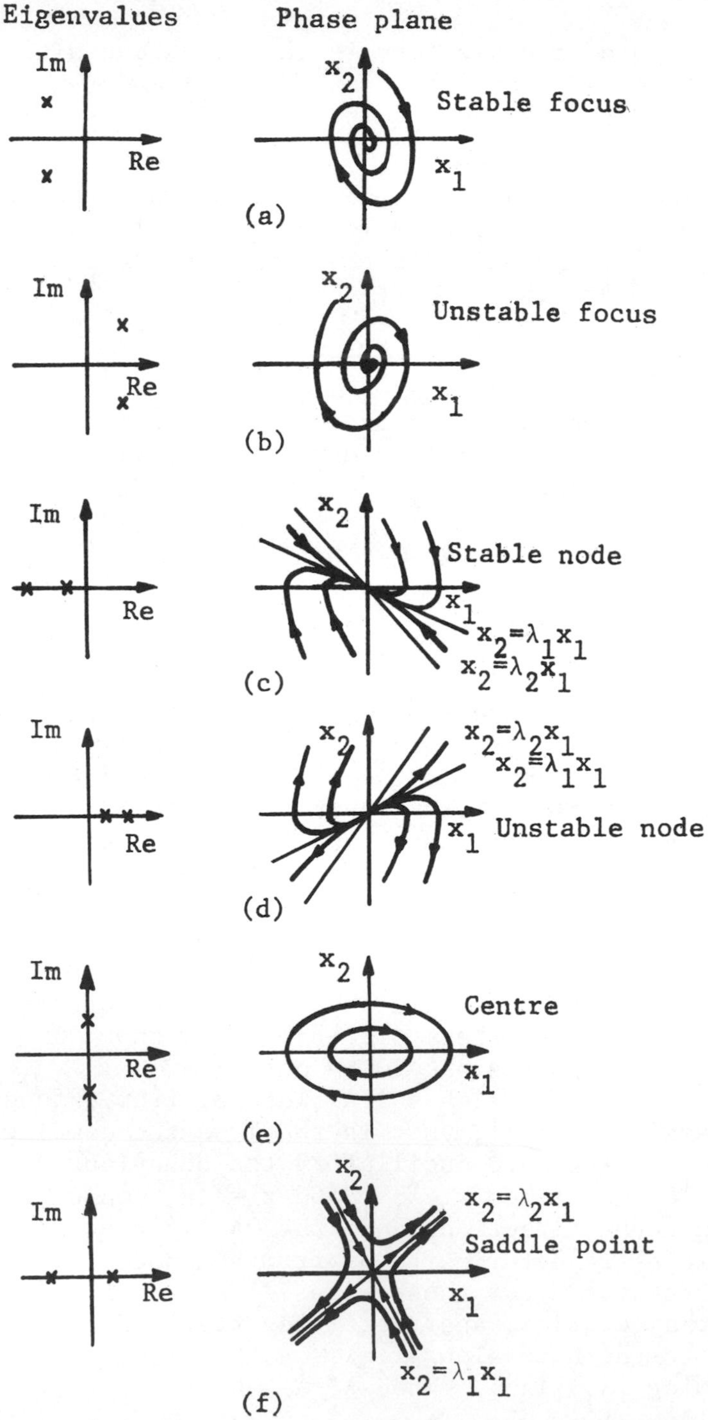

Figure 2.1 Phase portraits for type 0 system

TABLE 2.1

Types of Singular Point

Eigenvalues λ_1, λ_2	a_0, a_1	Singular Point
Complex Re(λ)<0	a_1 +ve, $a_1^2<4a_0$	Stable focus
Complex Re(λ)>0	a_1 −ve, $a_1^2<4a_0$	Unstable focus
λ_1, λ_2, real −ve	a_0 +ve, a_1 +ve, $a_1^2 \geq 4a_0$	Stable node
λ_1, λ_2, real +ve	a_0 +ve, a_1 −ve, $a_1^2 \geq 4a_0$	Unstable node
λ_1, λ_2, imaginary	$a_1 = 0$	Centre
λ_1 real −ve λ_2 real +ve	a_0 −ve	Saddle point

(c) The trajectories passing through the saddle point, which are the lines $x_2=\lambda_1 x_1$ and $x_2=\lambda_2 x_1$, are known as separatrices.

For the nonlinear system of eq. (2.8) singular points occur where $x_2=0$ and $f(x_1,0)=0$. Depending upon the form of $f(x_1,0)$ multiple singular points may exist but they will all lie on the x_1 axis. As mentioned earlier the type of singular point is found by linearising the nonlinear equation about the singular point, which gives

$$\dot{x}_2 = -(x_1 - x_s) \left.\frac{\partial f}{\partial x_1}\right|_{\substack{x_1=x_s \\ x_2=0}} - x_2 \left.\frac{\partial f}{\partial x_2}\right|_{\substack{x_1=x_s \\ x_2=0}}$$

so that

$$a_0 = \partial f/\partial x_1 \Big|_{\substack{x_1=x_s \\ x_2=0}} \tag{2.61}$$

$$a_1 = \partial f/\partial x_2 \Big|_{\substack{x_1=x_s \\ x_2=0}} \tag{2.62}$$

at the singular point $(x_s,0)$.

2.5.2 Limit cycles

A limit cycle is a steady state oscillation unique to a nonlinear system. It has a distinct geometric configuration on a phase plane portrait, namely that of an isolated closed path. Trajectories near to the limit cycle will either converge to it or diverge from it but cannot cross it. The limit cycle thus divides the phase plane into regions inside and outside the limit cycle. The closed curves corresponding to the second order system with no damping, shown about the centre in Fig. 2.1(e) are not limit cycles.

A limit cycle is called stable (unstable) if trajectories originating from inside or outside the limit cycle converge to it (diverge from it). It is called semi or half stable if trajectories from inside converge to it and those outside diverge from it or vice versa. Only a stable limit cycle will be observed in practice although it is possible to simulate unstable limit cycles in second order systems by simulating the system differential equation obtained by replacing t with $-\tau$. This corresponds to a reverse time situation so that time along the trajectories for the modified differential equation is in the reverse direction to that of the original equation.

Since a particular differential equation may have more than one limit cycle solution, limit cycles can be nested, that is they may exist one inside the other in a phase plane portrait. As pointed out in Chapter 1 the question of whether a system is stable often becomes one of whether the system has a limit cycle so that any criteria regarding conditions for the non existence of limit cycles can be valuable.

Two results of this type for second order systems have been known for many years. Probably the most useful is that due to Bendixson [12] which, for the system of eqs. (2.8) states that no limit cycle can exist in a region of the phase plane for which $\partial f/\partial x_2$ has an invariant sign and is not identically zero. An index theorem due to Poincaré states that the sum of the indices of singularities enclosed within a limit cycle must be +1. A saddle point has an index of -1 and other singularities an index of +1.

2.6 PHASE PLANE TRAJECTORIES

Before the advent of modern computational techniques, either analog or digital, the behaviour of many second order differential equations was studied by sketching their phase plane trajectories. Several graphical methods have been described in the literature [13] for doing this, as well as for determining the time taken to move along a particular trajectory. These approaches of which the most general is the method of isoclines, remain very useful for obtaining a quick qualitative estimate of the behaviour. The method of isoclines is based on eq. (2.57) which shows that the slope of a trajectory is equal to m along the curve

$$mx_2 + f(x_1,x_2) = 0. \tag{2.63}$$

In particular for the linear case this curve becomes the straight line

$$mx_2 + a_1x_2 + a_0x_1 = 0 \tag{2.64}$$

which has a slope $-a_0/(a_1+m)$ and passes through the origin. By selecting a few values of m the isoclines given by eq. (2.64) can be drawn. With arrows of slope m drawn across the isocline to indicate the trajectory slope a reasonable qualitative picture of a phase portrait can be obtained once a few isoclines are drawn. For the nonlinear situation the isoclines are the curves of eq. (2.63) and more effort is required to plot them.

To illustrate some of the points made in this and the previous section we consider briefly the van der

Pol equation

$$\ddot{x} + \mu(x^2 - 1)\dot{x} + x = 0 \qquad (2.65)$$

where μ is a positive constant. Here

$$f(x_1,x_2) = x_1 + \mu(x_1^2 - 1)\, x_2$$

so that the only singular point is the origin. Also on partial differentiation, according to eqs. (2.61) and (2.62), we obtain

$$a_0 = 1 \text{ and } a_1 = -\mu.$$

The origin is thus an unstable focus if $\mu<2$ and an unstable node for $\mu\geq 2$. From a physical viewpoint we observe that the damping factor $\mu(x_1^2-1)$ is negative for $|x_1|<1$, as expected because the origin is unstable and is positive for $|x_1|>1$. This suggests the system may have a limit cycle, an observation which is supported by Bendixson's theorem, since $\partial f/\partial x_2=\mu(x_1^2-1)$. This involves the same factor used in the physical argument and changes sign when x_1 becomes greater than unity.

Fig. 2.2 shows, for $\mu=1$, the phase plane portrait, obtained using the method of isoclines, of a trajectory originating from the unstable focus at the origin. A stable limit cycle is seen to exist which has an x_1 amplitude of approximately 2.0. For smaller values of μ the limit cycle becomes more circular and for larger values of μ attains much larger values of x_2 with regions where dx_2/dx_1 becomes large. The x_1 amplitude, however, does not change significantly.

Nonlinear elements encountered in control systems can rarely be expressed so that $f(x_1,x_2)$ is a simple function, say a few terms involving low order powers of x_1 and x_2. On the other hand it is often appropriate to approximate a nonlinearity by a linear segmented characteristic. Phase plane trajectories can easily be obtained for this situation since the problem reduces to one of using different linear differential equations to describe the motion of trajectories in several separate regions of the phase plane. The boundaries dividing the various regions

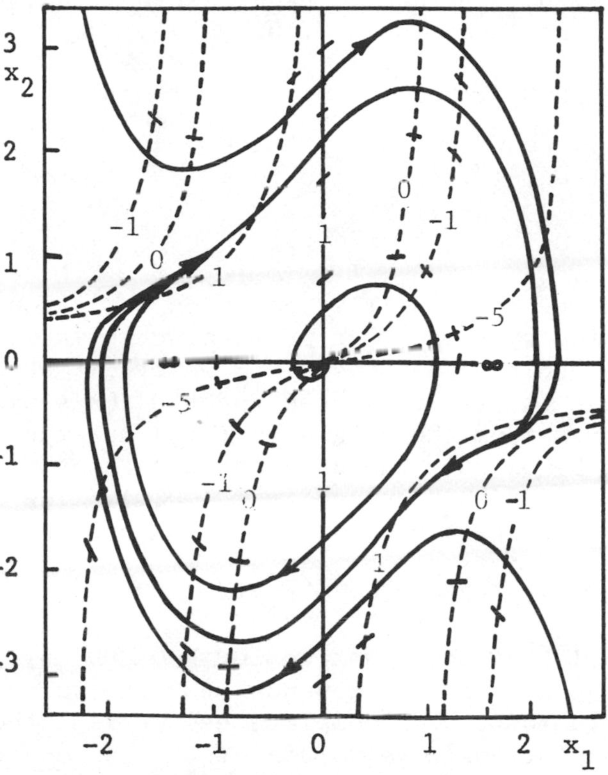

Figure 2.2 Phase plane for van der Pol equation

are easily evaluated and since they are also linear segmented they can easily be drawn on the phase plane. It may happen in this case that the singular point for trajectories in a particular region lies outside that region and therefore cannot be reached by the trajectories. Such a singular point is known as a virtual singular point [14]. Before considering applications of this approach a few additional points relating to the linear differential eq. (2.58) must be covered. We will refer to this equation, which apart from the gain, K, is the equivalent of the transfer function

$$\frac{X(s)}{U(s)} = \frac{K}{s^2 + a_1 s + a_0} \tag{2.66}$$

as being of type 0 when it has no integrations, that

is $a_0 \neq 0$, $a_1 \neq 0$; of type 1 when it has one integration, that is $a_0 = 0$, $a_1 \neq 0$ and of type 2 when it has two integrations, that is $a_0 = a_1 = 0$.

2.6.1 Type 0 system

Firstly, we write the differential equation equivalent of eq. (2.66) in the form

$$\ddot{x} + 2\zeta\omega_o\dot{x} + \omega_o^2 x = \omega_o^2 u(t) \tag{2.67}$$

which is the more familiar notation for second order engineering systems. Our prime concern is with initial condition responses. However, it should be noted that if u(t) is a step input U the singular point moves from the origin to (U,0) but the trajectory will be exactly the same shape as that for an initial condition response from (-U,0). The slope of a phase plane trajectory is given by

$$dx_2/dx_1 = (-2\zeta\omega_o x_2 - \omega_o^2 x_1)/x_2. \tag{2.68}$$

The solution of the equation for U=0 and initial conditions x_{10} and x_{20} depends upon the value of the damping ratio ζ. For completeness we list these solutions for positive values of ζ.

Zero damping, $\zeta=0$

Here

$$x_1 = x_{10} \cos \omega_o t + (x_{20}/\omega_o) \sin \omega_o t \tag{2.69}$$

$$x_2 = -x_{10}\, \omega_o \sin \omega_o t + x_{20} \cos \omega_o t. \tag{2.70}$$

The equation of a phase plane trajectory is easily shown by eliminating t from the above equations, or by integrating directly eq. (2.68), to be an ellipse, as shown in Fig. 2.1(e). It becomes a circle if $x_{2N} = x_2/\omega_o$ is plotted along the ordinate axis.

Light damping, $0<\zeta<1$

In this case

$$x_1 = \frac{e^{-\zeta\omega_0 t}}{(1-\zeta^2)^{1/2}} [x_{10} \sin\{(1-\zeta^2)^{1/2}\omega_o t + \phi\}$$

$$+ (x_{20}/\omega_o) \sin\{(1-\zeta^2)^{1/2}\omega_o t\}] , \quad (2.71)$$

$$x_2 = \frac{-e^{-\zeta\omega_0 t}}{(1-\zeta^2)^{1/2}} [\omega_o x_{10} \sin\{(1-\zeta^2)^{1/2}\omega_o t\}$$

$$+ x_{20} \sin\{(1-\zeta^2)^{1/2} \omega_0 t - \phi\}], \quad (2.72)$$

where $\phi = \cos^{-1}\zeta$.

When $x_{20} = 0$ it is easily shown that successive overshoots and undershoots in the oscillatory response of x_1 are reduced by Δ where

$$\Delta = \exp\{-\zeta\pi/(1-\zeta^2)^{1/2}\} \quad (2.73)$$

The phase plane portrait is shown in Fig. 2.1(a).

Critical damping, $\zeta=1$

In this case the solutions are

$$x_1 = e^{-\omega_0 t}\{x_{10}(1 + \omega_o t) + x_{20} t\} \quad (2.74)$$

$$x_2 = e^{-\omega_0 t}\{-x_{10} \omega_o^2 t + x_{20}(1 - \omega_o t)\} \quad (2.75)$$

The phase plane portrait has the single eigenvector given by

$$x_2 + \omega_o x_1 = 0 \quad (2.76)$$

Heavy damping, $\zeta>1$

The solutions are

$$x_1 = \frac{e^{-\zeta\omega_0 t}}{(\zeta^2-1)^{1/2}} [x_{10} \sinh\{(\zeta^2-1)^{1/2}\omega_o t + \phi\}$$

$$+ (x_{20}/\omega_o) \sinh\{(\zeta^2-1)^{1/2}\omega_o t\}] , \quad (2.77)$$

$$x_2 = \frac{e^{-\zeta\omega_o t}}{(\zeta^2 - 1)^{1/2}} [\omega_o x_{10} \sinh \{(\zeta^2 - 1)^{1/2}\omega_o t\}$$

$$+ x_{20} \sinh \{(\zeta^2 - 1)^{1/2}\omega_o t - \phi\}] , \qquad (2.78)$$

where $\phi = \cosh^{-1}\zeta$.

The phase portrait is as shown in Fig. 2.1(c) with the two eigenvectors given by

$$x_2 - \omega_o \{-\zeta \pm(\zeta^2 - 1)^{1/2}\} x_1 = 0. \qquad (2.79)$$

For other than $\zeta=0$ the equations for the phase plane trajectories in cartesian coordinates are quite involved and therefore of minor practical use.

2.6.2 Type 1 system

Here we consider the differential equation equivalent of eq. (2.66) in the form

$$\ddot{x} + \alpha\dot{x} = KU \qquad (2.80)$$

where U is a constant input. The slope of a trajectory is

$$dx_2/dx_1 = (-\alpha x_2 + KU)/x_2 . \qquad (2.81)$$

This equation has no singular points unless U=0, in which case all points on the x_1 axis are singular and all the phase plane trajectories are straight lines of slope $-\alpha$ directed to these singular points. From eq. (2.81) isoclines of slope m are the lines

$$x_2 = KU/(m + \alpha) \qquad (2.82)$$

which are parallel to the x_1 axis. The isocline with zero slope, that is $x_2=KU/\alpha$, is also a trajectory so that all trajectories tend to this line asymptotically. In addition all trajectories cut the x_2 axis orthogonally. Integration of eq. (2.81) gives the equation

$$x_1 - x_{10} = \{-(x_2-x_{20})/\alpha\} + (KU/\alpha^2)\log_e\{(\alpha x_{20}-KU)$$

$$/(\alpha x_2 - KU)\} \qquad (2.83)$$

for a phase plane trajectory. A sketch of the trajectories for U positive (in full) and U negative (dotted) is shown in Fig. 2.3.

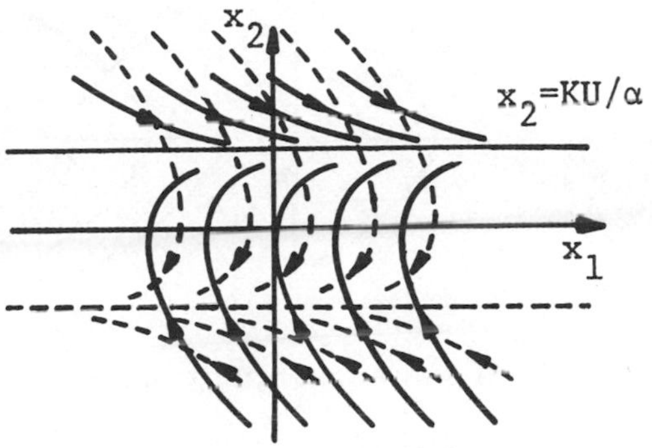

Figure 2.3 Phase plane trajectories for type 1 system

2.6.3 <u>Type 2 system</u>

Here we consider the differential equation

$$\ddot{x} = KU \tag{2.84}$$

which describes motion with constant acceleration if x is assumed to be a distance. The slope of a phase plane trajectory is

$$dx_2/dx_1 = KU/x_2 \tag{2.85}$$

which on integration gives

$$x_1 - x_{10} = (1/2KU)(x_2^2 - x_{20}^2). \tag{2.86}$$

For KU≠0 this is the equation of a parabola and the phase plane trajectories are shown in Fig. 2.4. Eq. (2.85) again has no singular points unless KU=0, in which case all points on the x_1 axis are singular and the phase plane trajectories are straight lines parallel to the x_1 axis, moving to the right for $x_2>0$ and to the left for $x_2<0$.

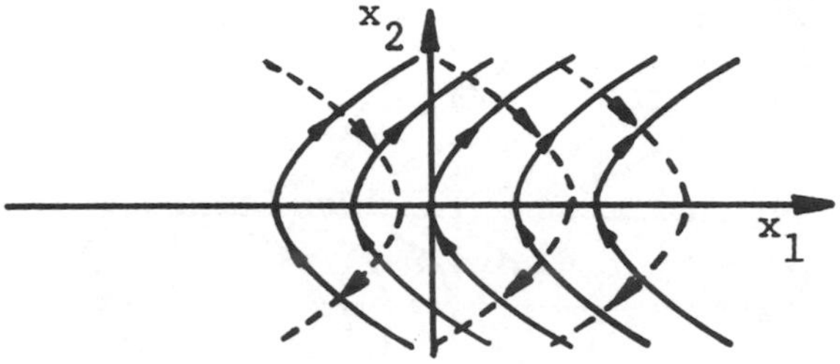

Figure 2.4 Phase plane trajectories for type 2 system

2.7 CONTROL APPLICATIONS

To illustrate the applicability of phase plane methods to the determination of stability and performance of typical control loops several examples are considered below. These have been selected to illustrate, amongst other aspects, that more than one nonlinear element can be handled by the phase plane approach; that for relay systems the method can be used if there is also a time delay and that a sliding state in relay systems, a situation where the relay switches rapidly on and off, may occur.

Example 1

Investigate the stability of a feedback system containing the relay with dead zone and hysteresis shown in Fig. 2.5 for K=1, δ=1 and Δ=0.5 and the linear transfer function G(s)=20/s(s+1).

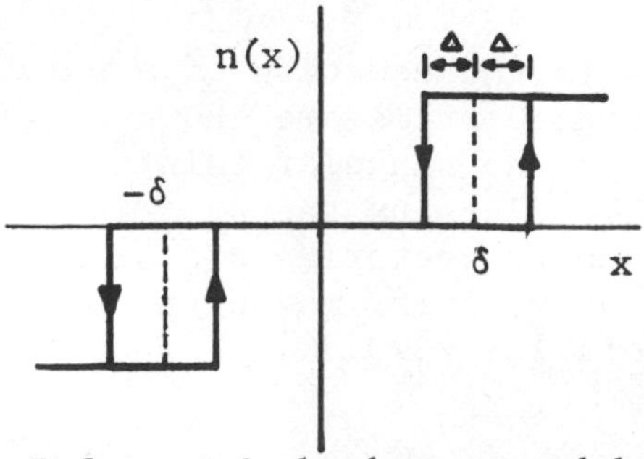

Figure 2.5 Relay with dead zone and hysteresis

Since the relay switches on when its input reaches 1.5 and off when it reaches 0.5 the switching boundaries are as shown in Fig. 2.6. Also as the only relay output values are $\pm$ 1 and 0 the phase portraits in the regions correspond to those of a type 1 system with $U=\pm 1$ and 0. A trajectory starting at point $B(-(\delta-\Delta),y_2)$ will reach point $C(\delta+\Delta,y_1)$, since the motion is linear, where

$$y_1 - y_2 = 2\delta . \tag{2.87}$$

Thus, for a limit cycle to occur, as shown in Fig. 2.6 a trajectory starting at $A(-(\delta+\Delta),-y_1)$ must pass through B. Using eq. (2.83) the equation of this trajectory is

$$-(\delta-\Delta) + (\delta+\Delta) = -(y_2+y_1) + K \log_e\{(K+y_1)/(K-y_2)\}$$

which on substituting for the parameters, K, δ and Δ and y_1 from eq. (2.87) gives

$$(2y_2-1)/20 = \log_e \{(18+y_2)/(20-y_2)\} .$$

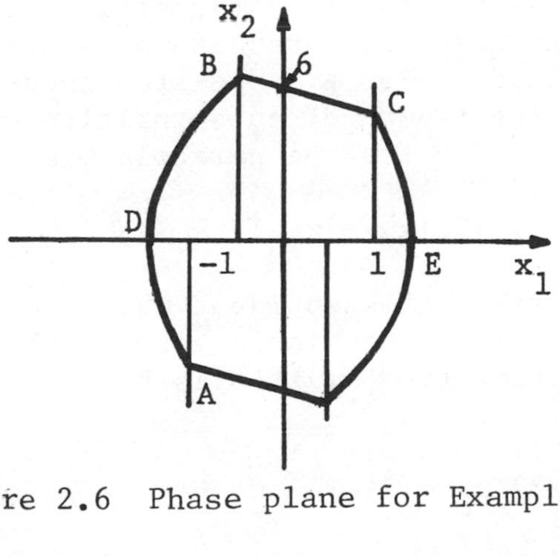

Figure 2.6 Phase plane for Example 1

The solution for y_2 is 6.75 so that a limit cycle exists as sketched in Fig. 2.6. Substituting the value of y_2 in the trajectory equation shows that the point D is (-1.48,0). By drawing trajectories near to the limit cycle it is found that those both inside and outside converge to it, indicating that the limit cycle is stable.

Example 2

Fig. 2.7 shows the block diagram of a control system. We wish to investigate the response of the system to a large step input and one of value 10.

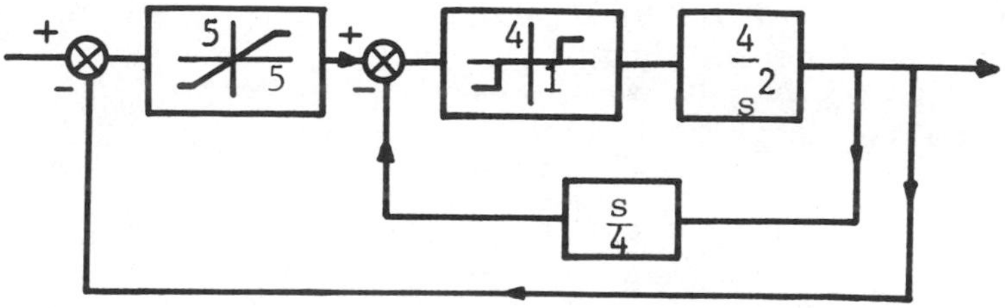

Figure 2.7 Block diagram for Example 2

The input to the relay is $-f(x_1)-x_2/4$, where $f(x_1)$ denotes the saturation characteristic. The boundaries of the relay dead band operation are $-x_1-(x_2/4)=\pm 1$ with no saturation and $\pm 5-(x_2/4)=\pm 1$ with saturation, which are shown on Fig. 2.8. Within the relay dead band the trajectories are horizontal lines and outside the dead band they are parabolae as shown in the previous section. For a large step input all trajectories reach the region of no saturation with $x_2=16$. Since the equation for the parabolae is

$$x_2^2 - x_{20}^2 = \pm 32(x_1 - x_{10})$$

the minimum value of step input which reaches this level is 13 units. Continuing around this large step input trajectory from A, if B is the point $(x_b,0)$ then,

$$-16^2 = -32(x_b + 3)$$

so that $x_b=5$. If the coordinates of C are (x_c,y_c) then

$$-x_c - (y_c/4) = -1$$

and

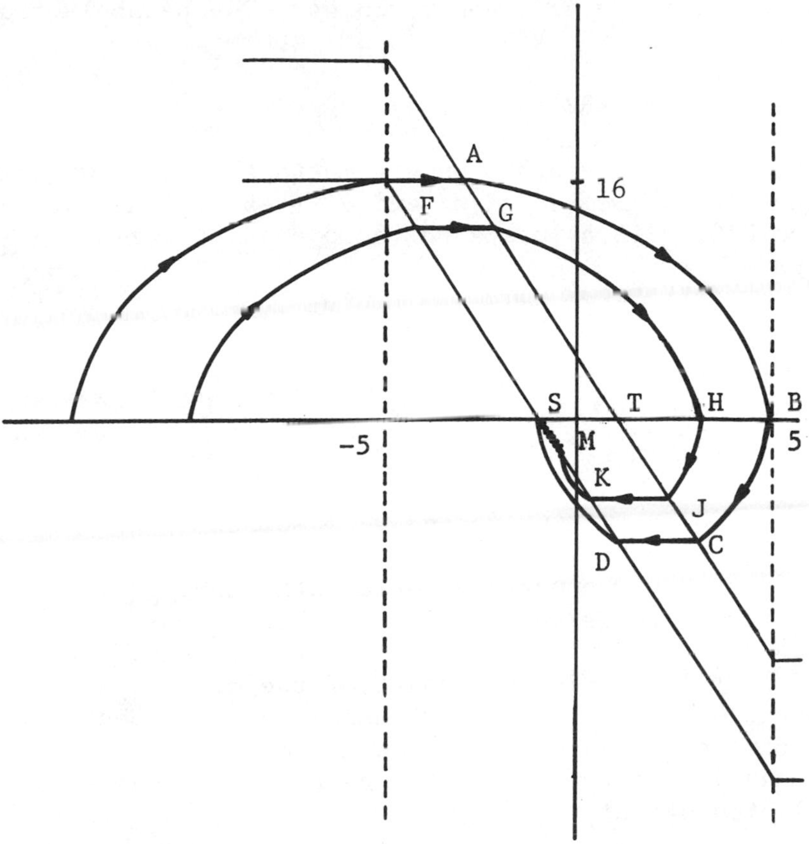

Figure 2.8 Phase plane for Example 2

$$-y_c^2 = -32(x_c - 5)$$

giving $x_c=3$, $y_c=-8$. By symmetry the motion from D is to the edge of the dead band at S where the motion stops since the line ST is a set of singular points.

Considering now the input step of 10 units, the relay switches into the dead band at point $F(x_f,y_f)$ where

$$-x_f - (y_f/4) = 1$$

and

$$y_f^2 = 32(x_f + 10)$$

which gives x_f=-4.33 and y_f=13.44. The parabola from G crosses the x_1 axis at $H(x_h,0)$ where

$$-13.44^2 = -32(x_h + 2.33)$$

which gives $x_h = 3.32$, and it meets the next switching line at J where $x_j = 2.37$ and $y_j = -5.5$. The parabola from K(0.37,-5.5) meets the same switching line again at M where

$$-x_m - (y_m/4) = 1$$

and

$$y_m^2 - (5.5)^2 = 32(x_m - 0.37)$$

which yields $y_m = -2.45$ and $x_m = -0.36$.

The motion has now reached a portion of the switching boundary where trajectories at either side are directed towards the boundary. The motion from M is thus down the switching boundary to the singular point at S. During this part of the motion the relay output switches rapidly between +1 and 0. Because of the manner in which the motion moves along the boundary this type of system behaviour is called a 'sliding state'.

Example 3

Here we consider the relay of Fig. 2.5 with Δ=0, δ=1 and h=1 driving a plant of transfer function $G(s)=K/s^2$. The feedback signal at the relay input is $-x_1' - \lambda x_2$ where $x_1'(t)=x_1(t-T_d)$, the output delayed by T_d secs.

The relay switches when

$$-x_1' - \lambda x_2 = \pm 1,$$

and in order to draw these boundaries in the x_2-x_1 phase plane x_1' must be expressed in terms of x_1. Since x_1' is equal to the value of x_1 T_d seconds previously the relationship between x_1 and x_1' depends upon the particular motion. When motion is at a constant velocity for relay dead band operation

$x_1 - x_1' = x_2 T_d$. This motion occurs for $x_2>0$ before the switching line $-x_1' - \lambda x_2 = -1$, so that the equation of the switching curve in the x_1-x_2 plane is

$$x_2 T_d - x_1 - \lambda x_2 = -1$$

that is

$$-x_1 - x_2(\lambda - T_d) = -1 \qquad (2.87)$$

a line of slope $-1/(\lambda-T_d)$. Before the switching line $-x_1' - \lambda x_2=1$ for $x_2>0$ the motion is along the parabola

$$x_2^2 - x_{20}^2 = 2K(x_1 - x_{10})$$

where (x_{10},x_{20}) is an initial condition and (x_1,x_2) is the point reached after time t. We also have

$$x_1 - x_{10} = x_{20}\, t + Kt^2/2$$

and after time $t-T_d$, when $x_1=x_1'$

$$x_1' - x_{10} = x_{20}(t - T_d) + K(t - T_d)^2/2.$$

Subtracting these equations gives

$$x_1 - x_1' = x_{20}T_d + Kt\, T_d - KT^2{}_d/2 .$$

In addition, however,

$$x_2 = x_{20} + Kt$$

so eliminating x_{20} gives

$$x_1 - x_1' = x_2 T_d - KT_d^2/2 .$$

Substituting this value in the switching line equation yields

$$-x_1 + x_2 T_d - (KT_d^2/2) - \lambda x_2 = 1$$

that is

$$-x_1 - x_2(\lambda - T_d) = 1 + KT_d^2/2 . \qquad (2.88)$$

The two switching lines given by eqs. (2.87) and (2.88) are shown in Fig. 2.9 for $T_d > \lambda$. It is seen from the figure that a limit cycle can exist passing through the points ABCD, when the ordinate axis bisects AD and BC. Sketching of phase plane trajectories near to this limit cycle shows that it is unstable with trajectories both inside and outside diverging from it. Further it can be seen that for $\lambda > T_d$ no limit cycle can exist and the system is stable. If the coordinates of B and C are $(-x_b, y_b)$ and (x_b, y_b) respectively it is easily shown from eqs. (2.87) and (2.88) that

$$x_b = 1 + (KT_d^2/4) \text{ and } y_b = KT_d^2/4(T_d - \lambda).$$

By calculating the time taken on the portion ABCD of the limit cycle its period is found to be

$$\frac{T_d^2}{T_d - \lambda} + \frac{4(4 + KT_d^2)(T_d - \lambda)}{KT_d^2} .$$

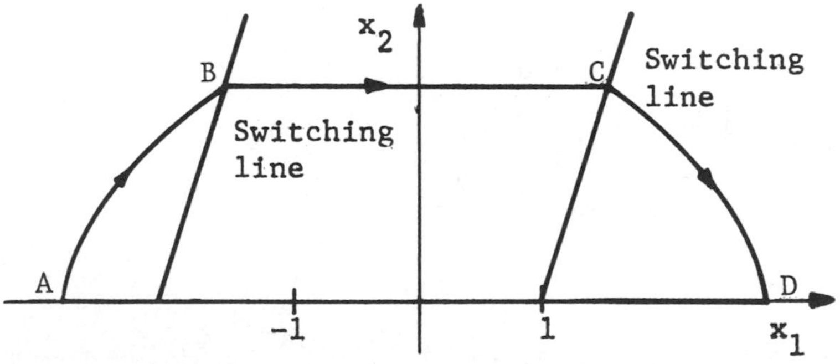

Figure 2.9 Phase plane for Example 3

2.8 SUMMARY

In this chapter some analytical methods and the phase plane approach for the study of nonlinear systems have been presented. Although the methods are primarily restricted to second order systems there are many control applications where they can be used to advantage. Most of the methods can be extended to deal with forced oscillations which are not considered here. The phase plane approach, in particular, tends to complement the DF method presented in Chapter 4 as the DF procedure usually gives more accurate results for higher order systems.

Many texts have been written on second order systems mainly considering polynomial rather than linear segmented nonlinearities. The books by Hayashi [15], Minorsky [16], Cunningham [17], Ku [18] and Andronov, Vitt and Khaikin [19] are particularly recommended for a deeper study of these topics as is the special issue of the IRE Transactions [20] in memory of van der Pol. Several control systems texts, which deal with nonlinear problems, have a chapter or more on the phase plane approach, these include the books by West [21], Gibson [22], Graham and McRuer [23], Atherton [13] and Ogata [24].

REFERENCES

1. Poincaré, H.: "Memoire sur les courbes definies par une equation differentielle", J. de Mathematiques, 7, 1881, pp. 375-422, and 8, 1882, pp. 251-296.

2. Rayleigh, Lord: "On maintained vibrations", Phil. Mag., Vol. 15, No. 94, Series 5, 1883, pp. 229-235.

3. Van der Pol, B.: "On relaxation oscillations", Phil. Mag., 7th Series, Vol. 2, pp. 978-992, 1926.

4. Volterra, V.: "Lecons sur la théorie mathematique de la lutte pour la vie", Gauthier-Villars, Paris, 1931.

5. Barnett, S.: "Introduction to mathematical control theory", Clarendon Press, Oxford, 1975.

6. Cesari, L.: "Asymptotic behaviour and stability problems in ordinary differential equations", Academic Press, Inc., 1963.

7. Krylov, N. and Bogolyubov, N.: "Introduction to nonlinear mechanics", Princeton University Press, Princeton, N.J., 1943.

8. Barkham, P.G.D. and Soudack, A.C.: "An extension to the method of Kryloff and Bogoliuboff", Int. J. Control, Vol. 10, pp. 377-392, 1969.

9. Barkham, P.G.D. and Soudack, A.C.: "Limit-cycle phenomena in non-linear time delay systems", Int. J. Control, Vol. 27, pp. 407-420, 1978.

10. Halanay, A.: "Differential equations, stability, oscillations, time lags", Academic Press, 1966.

11. Luus, R. and Lapidus, L.: "An averaging technique for stability analysis", Chemical Eng. Science, Vol. 21, pp. 159-181, 1966.

12. Bendixson, I.: "Sur les courbes definies par des equations differentielles", Acta Mathematica, Vol. 24, pp. 1-88, 1901.

13. Atherton, D.P.: "Nonlinear control engineering", Van Nostrand Reinhold, London, 1975, Chapter 2.

14. Kalman, R.E.: "Phase-plane analysis of automatic control systems with nonlinear gain elements", Trans. AIEE, Vol. 73, Pt. II, pp. 383-390, 1954.

15. Hayashi, C.: "Nonlinear oscillations in physical systems", McGraw-Hill, New York, 1964.

16. Minorsky, N.; "Nonlinear oscillations", D. Van Nostrand, Princeton, N.J., 1962.

17. Cunningham, W.J.: "Introduction to nonlinear analysis", McGraw-Hill, New York, 1958.

18. Ku, Y.H.: "Analysis and control of nonlinear systems", Ronald Press Co., New York, 1958.

19. Andronov, A.A., Vitt, A.A. and Khaikin, S.E.: "Theory of oscillators", Addison-Wesley, Reading, Mass., 1966.

20. van der Pol, B.: "Special issue in memory of Dr. Balth van der Pol", IRE Trans. CT-7, No. 4, 1960.

21. West, J.C.:"Analytical techniques for nonlinear control systems", English University Press, London, 1960.

22. Gibson, J.F.: "Nonlinear automatic control", McGraw-Hill, New York, 1963.

23. Graham, D. and McRuer, D.: "Analysis of nonlinear systems", Wiley, New York, 1961.

24. Ogata, K.: "Modern control engineering", Prentice Hall, New Jersey, 1970, Chapter 12.

PROBLEMS

1. Show that for small α the perturbation method to order α^2 gives a solution to the van der Pol eq. (2.51) of $\omega = 1 - (\alpha^2/16)$ and

$$x = \{2 - (\alpha^2/8)\} \cos \omega t + (3\alpha/4) \sin \omega t +$$

$$(3\alpha^2/16) \cos 3\omega t - (\alpha/4) \sin 3\omega t -$$

$$(5\alpha^2/96) \cos 5\omega t.$$

2. Show that for a differential equation of the form $\ddot{x} + f(\dot{x}) + x = 0$ the slope of a trajectory at any point P in the phase plane is a line perpendicular to SP where S is obtained by (i) plotting the curve $x + f(\dot{x}) = 0$, (ii) drawing a line from P parallel to the x axis to meet the curve drawn in (i) at R and (iii) drawing a line from R parallel to the y axis to meet the x axis at S.

3. Use the method of problem 2 to sketch phase plane trajectories for $\ddot{x} + \dot{x}|\dot{x}| + x = 0$ and the Raleigh eq. (2.17) with $\mu = 1$. Estimate the limit cycle in the latter case.

4. Show that by a change of variables the van der Pol eq. (2.51) can be put in the form required to apply the construction (Liénard construction) of problem 2. Estimate the limit cycle for $\mu = 1$.

5. Show that for $\zeta > 1$ the lines $x_2 = \lambda x_1$, where $\lambda = (-\zeta \pm \sqrt{\zeta^2 - 1})\omega_0$ are eigenvectors of $\ddot{x} + 2\zeta\omega_0\dot{x} + \omega_0^2 x = 0$.

6. Determine the ratio of the times to move down the eigenvectors of problem 5 to the origin from points with equal abscissae.

7. Evaluate eqs. (2.71) and (2.72) and obtain the value given for Δ in eq. (2.73).

8. For the conditions of problem 5, what is the maximum value of the ratio y/x if a trajectory starting from the point P(-x,y) in the phase plane must not overshoot the origin?

9. For the differential equation $\ddot{x} + f(x) = 0$ evaluate the period of an oscillation about the centre at the origin if (i) $f(x) = x$, (ii) $f(x) = h \operatorname{sgn} x$ and (iii) $f(x) = \sin x$.

10. A relay with $\Delta = \delta = 0$ and $h = 1$ provides the input to a transfer function $G(s) = 10/s(s+\alpha)$ whose output is x. The input signal to the relay is $-x + \dot{x}$, find the value of α, say α_c, for the system to be stable. If α is less than α_c does the system have a limit cycle; if so evaluate its amplitude and frequency.

11. Repeat problem 10 if the relay input is $-x + \dot{x} - \dot{x}^3/3$.

12. A relay with $\delta = 1$, $\Delta = 0.5$ and $h = 1$ drives a process with transfer function $10e^{-sT}/s$. Use the phase plane approach to evaluate the maximum value of T for which the system will be stable.

CHAPTER 3

Absolute Stability Criteria

3.1 INTRODUCTION

Many distinguished researchers have made contributions to the literature on methods for determining the stability of the nonlinear feedback system of Fig. 1.1. Unlike the linear situation, however, where the algebraic procedures of the Hurwitz-Routh criterion or the graphical method of the Nyquist criterion provide necessary and sufficient conditions for stability, the nonlinear problem is not completely solved. In fact, before the relatively recent development of frequency domain criteria no truly satisfactory approach was available which could be easily used by the non-mathematician. The various frequency domain criteria, which provide sufficient, but not necessary conditions for stability, often produce conservative results because of the generality of the approach which uses a relatively small amount of information regarding the nonlinearity.

The aim of this chapter is to present what are believed to be the most useful absolute stability techniques for investigating the stability of nonlinear feedback systems. We begin by considering the Liapunov [1] approach before proceeding to the frequency domain results. Liapunov's work is valuable not only because there are still situations where the method can prove advantageous but also because most of the frequency domain results, including the original result by Popov [2] have been obtained using the Liapunov method. Since the primary concern of this monograph is to present and compare various techniques for stability investigation the proofs of the frequency domain criteria are not given here. The interested reader will find proofs of the theorems in references [3-5] and alternative proofs obtained somewhat more recently using functional analysis methods in references [5-7]. Section 3.7 includes some results on criteria for the absence of limit cycles which again can be proved using functional analysis. Finally, we take a brief look at criteria both for the existence of limit cycles and for instability.

3.2 THE LIAPUNOV METHOD

Liapunov, a Russian mathematician, presented the fundamental concepts of the stability theory named after him in 1892. This theory received little attention, especially outside the USSR, until around 1950. The publication of the work of Lur'e [8] and the book by LaSalle and Lefschetz [9] brought the work to the attention of control engineers. Many refinements have been added since then, but the major problem of finding a suitable Liapunov function, $V(x)$, for a given nonlinear system still remains.

The direct or second method of Liapunov, is a time domain method based on a nonlinear state space description of the system. In our case, since we are restricting attention to the time invariant case, the problem becomes one of determining the stability of an equilibrium state, which without loss of generality will be assumed to be the origin of the state space of the nonlinear system

$$\dot{x} = f(x). \qquad (3.1)$$

Liapunov's theorems on the stability and asymptotic stability of eq. (3.1) are as follows.

Theorem 1

The null solution, or the equilibrium state at the origin of the system described by eq. (3.1) is stable if in a small neighbourhood, R, around the origin there exists a positive definite function $V(x)$ such that its derivative $\dot{V}(x)$ is negative semi-definite in that region.

Theorem 2

The null solution of eq. (3.1) is asymptotically stable if in a small neighbourhood, R, around the origin there exists a positive definite function V(x) such that its derivative $\dot{V}(x)$ is negative definite in R.

A theorem due to LaSalle, which follows, enables the conditions on $\dot{V}(x)$ in the above theorem to be relaxed.

Theorem 3

The null solution of eq. (3.1) is asymptotically stable if, in a small neighbourhood, R, around the origin, there exists a positive definite function V(x) such that its derivative $\dot{V}(x)$ is negative semi-definite in R and is non-zero in R along any solution of eq. (3.1) other than the null solution.

If the conditions of theorem 3 are satisfied for all x then the equilibrium state of eq. (3.1) is asymptotically stable in the large. A function V(x) which satisfies the conditions of any of the above theorems is known as a Liapunov function. Since the above theorems provide only sufficient conditions for stability of the equilibrium state, failure to find a Liapunov function does not prove instability but simply a failure to prove stability. It may also happen that the conditions required to prove stability for a particular choice of Liapunov function are very conservative.

The above theorems require V(x) to be a positive definite function the definition of which is as follows: -

The function V(x) is positive definite (or semi-definite) in S if, for all x in S

(i) V(x) has continuous partial derivatives with respect to the components of the vector x

(ii) $V(0) = 0$

(iii) $V(x) > 0$ for $x \neq 0$ (or $V(x) \geq 0$).

If the inequality signs in (iii) are reversed then the conditions for a negative definite (or semi-definite) function are obtained.

Consider the function

$$V(x) = x_1^2 + x_2^2 + x_3^2$$

which is zero only when $x_1=x_2=x_3=0$. It is therefore positive definite in a three dimensional space but only positive semi-definite in spaces of more dimensions.

On the other hand

$$V(x) = (x_1 + x_2)^2$$

is only positive semi-definite in two space, since it is zero everywhere along the line $x_1 + x_2 = 0$.

The quadratic form scalar function

$$P(x) = x^T P x \qquad (3.2)$$

where

$$P = \begin{pmatrix} p_{11} & p_{12} & \cdots & p_{1n} \\ p_{21} & p_{22} & \cdots & p_{2n} \\ \vdots & \vdots & \ddots & \\ p_{n1} & p_{n2} & \cdots & p_{nn} \end{pmatrix}$$

is often used. It may also be written

$$P(x) = \sum_{i=j}^{n} \sum_{j=1}^{n} p_{ij} x_i x_j$$

from which it can be seen that the same P(x) can be realised by all choices of P for which $p_{ij} + p_{ji}$ are the same. P is therefore normally chosen to be symmetric, that is $p_{ij} = p_{ji}$, in which case the sign definiteness of P(x) can be obtained from the coefficients of P.

Sylvester's Theorem

A necessary and sufficient condition for the quadratic

form of eq. (3.2) with P symmetric, to be positive definite, is that each of the following determinants, from the leading principal minors of P, be positive:

$$\det(p_{11});\ \det\begin{pmatrix} p_{11} & p_{12} \\ p_{12} & p_{22} \end{pmatrix};\det\begin{pmatrix} p_{11} & p_{12} & p_{13} \\ p_{12} & p_{22} & p_{23} \\ p_{13} & p_{23} & p_{33} \end{pmatrix};..;\det(P).$$

P(x) is positive semi-definite if the determinants of all the principal minors, not just the leading ones, are non-negative. P is negative definite if − P is positive definite. Alternatively as the above P matrix is symmetric, and therefore Hermitian since its coefficients are real, it will have real eigenvalues. These must all be positive for P positive definite.

3.2.1 Application to linear systems

Liapunov's method is easily applied to find the stability of a linear system. Consider the linear autonomous system

$$\dot{x} = Ax \tag{3.3}$$

and choose

$$V(x) = x^T Px . \tag{3.4}$$

Differentiating and substituting for $\dot{x}$ and $\dot{x}^T$ gives

$$\dot{V}(x) = x^T\{A^T P + PA\}x .$$

Since we require $\dot{V}(x)$ to be negative definite we can write

$$\dot{V}(x) = -x^T Qx \tag{3.5}$$

so that

$$-Q = A^T P + PA. \tag{3.6}$$

Asymptotic stability of the linear system will thus be assured if with Q a chosen positive definite symmetric matrix and A the given system matrix, a symmetric positive definite matrix P can be found. Since the matrix equation is equivalent to a set of $n(n + 1)/2$ linear equations in $n(n + 1)/2$ unknowns for an nth order system, given Q one can in principle solve for P. Two theorems on the matrix eq. (3.6), known as the Liapunov equation, are particularly important.

Theorem 4

If $\lambda_1, \lambda_2 \ldots \lambda_n$ are the eigenvalues of the matrix A, then eq. (3.6) has a unique solution P if and only if $\lambda_i + \lambda_j \neq 0$ for all $i, j = 1, 2, \ldots n$.

Theorem 5

With Q a positive definite symmetric matrix, the solution P of eq. (3.6)

(a) is positive definite if A has only eigenvalues with negative real parts,

(b) is negative definite if A has only eigenvalues with positive real parts, and

(c) is indefinite if A has some eigenvalues with both positive and negative real parts.

A suitable Liapunov function $V(x) = x^T P x$ for a linear system can thus be found from the solution of eq. (3.6) with Q taken equal to I. Consider, for example, the second order system with

$$A = \begin{pmatrix} 0 & 1 \\ -2 & -3 \end{pmatrix} .$$

Let

$$P = \begin{pmatrix} a & b \\ b & c \end{pmatrix} ,$$

then substituting in eqs. (3.6) with $Q = I$ gives

$$\begin{pmatrix} 0 & -2 \\ 1 & -3 \end{pmatrix}\begin{pmatrix} a & b \\ b & c \end{pmatrix} + \begin{pmatrix} a & b \\ b & c \end{pmatrix}\begin{pmatrix} 0 & 1 \\ -2 & -3 \end{pmatrix} = \begin{pmatrix} -1 & 0 \\ 0 & -1 \end{pmatrix},$$

which has the solution $a = 5/4$, $b = c = 1/4$. Thus the Liapunov function is

$$V(x) = (1/4)\,\{5x_1^2 + 2x_1x_2 + x_2^2\}\,. \tag{3.7}$$

On the other hand if we had simply assumed any positive definite quadratic form for $V(x)$ it may not have been possible to prove stability. To illustrate this, consider

$$V(x) = x_1^2 + x_2^2\,, \tag{3.8}$$

then

$$\begin{aligned} \dot{V}(x) &= 2x_1\dot{x}_1 + 2x_2\dot{x}_2 && (3.9) \\ &= 2x_1x_2 + 2x_2(-2x_1 - 3x_2) = -(2x_1x_2 + 6x_2^2), \end{aligned}$$

which is not negative definite and no conclusion can be drawn about stability.

If the elliptical contours for constant values of V in eq. (3.7) are drawn on a phase portrait for the system with the above A matrix, it will be found that all trajectories cut a contour only once. This then shows the physical interpretation of a Liapunov function; since it has a $\dot{V}(x)$ which is negative all trajectories move towards the singular point through successively decreasing values of V. On the other hand if the circles of eq. (3.8) are drawn for constant values of V a trajectory may cut each one more than once.

3.2.2 Application to nonlinear systems

A systematic procedure for finding a Liapunov function for a linear system has been presented in the previous section which starts by assuming a value for $\dot{V}(x)$. Unfortunately, no general approach exists for nonlinear systems. This is not surprising since to prove stability by Liapunov's method we are trying to find the equations in n space of surfaces near to the singular point which all trajectories cut from outside to inside. We have seen in the previous chapter that for nonlinear second order systems the shapes of these trajectories can change quite rapidly near to a singular point. Two possible approaches for finding Liapunov functions are given below. The first, the variable gradient method [10], is analogous to the linear approach in that a form for $\dot{V}(x)$ is assumed and the second, due to Lur'e, presents a useful form of Liapunov function. The proof by Popov of his frequency domain criterion explained in Section 3.4 used a slightly modified version of the Lur'e V(x) function [3].

Variable gradient method

Consider the general function V(x), then its derivative V(x) is given by

$$V(x) = \frac{\partial V}{\partial x_1}\frac{dx_1}{dt} + \frac{\partial V}{\partial x_2}\frac{dx_2}{dt} + \ldots + \frac{\partial V}{\partial x_n}\frac{dx_n}{dt}$$

$$= \text{grad } V(x) \,.\, \dot{x}$$

which on substituting from eq. (3.1) gives

$$\dot{V}(x) = g(x) \,.\, f(x)\,, \tag{3.10}$$

where . denotes a vector dot product and g(x) = grad V(x). Use is then made of the following.

Lemma

A necessary and sufficient condition for a continuous vector function g(x) to be the gradient of a scalar function is that the matrix

$$M = \begin{pmatrix} \partial g_1/\partial x_1 & \partial g_2/\partial x_1 & \ldots & \partial g_n/\partial x_1 \\ \partial g_1/\partial x_2 & \partial g_2/\partial x_2 & \ldots & \partial g_n/\partial x_2 \\ \vdots & \vdots & & \vdots \\ \partial g_1/\partial x_n & \partial g_2/\partial x_n & & \partial g_n/\partial x_n \end{pmatrix} \tag{3.11}$$

be symmetric where $g_1, g_2 \ldots g_n$ denote the components of the vector function $g(x)$.

The function $V(x)$ can be computed from

$$V(x) = \int_0^x g(x)\, dx$$

along any path joining the origin of the state space to the point x. Usually this is done along the axes, that is

$$V(x) = \int_0^{x_1} g_1(x_1, 0 \ldots 0)\, dx_1 + \int_0^{x_2} g_2(x_1, x_2, 0 \ldots 0)\, dx_2 + \ldots + \int_0^{x_n} g_n(x_1, x_2, \ldots x_n)\, dx_n . \quad (3.12)$$

The method then consists of selecting a vector function $g(x)$ with some adjustable parameters such that the matrix M of eq. (3.11) is symmetric and $V(x)$ is a Liapunov function; that is $\dot{V}(x)$ of eq. (3.10) is negative definite and $V(x)$ of eq. (3.12) positive definite.

Example

Consider a nonlinear second order system given by

$$\dot{x}_1 = x_2$$

$$\dot{x}_2 = -\alpha x_2 - x_1 - x_1^3 \quad (3.13)$$

and take

$$g(x) = \begin{pmatrix} ax_1 + bx_2 \\ cx_1 + dx_2 \end{pmatrix},$$

where a, b, c, and d may be functions of x_1 and x_2. The matrix M is symmetric for all x_1 and x_2 if

$$\partial g_1/\partial x_2 = \partial g_2/\partial x_1 ,$$

that is

$$x_1 \frac{\partial a}{\partial x_2} + b + x_2 \frac{\partial b}{\partial x_2} = c + x_1 \frac{\partial c}{\partial x_1} + x_2 \frac{\partial d}{\partial x_1} . \quad (3.14)$$

Using eq. (3.10)

$$\dot{V}(x) = (ax_1 + bx_2)x_2 + (cx_1 + dx_2)(-\alpha x_2 - x_1 - x_1^3)$$
$$= x_1x_2\ (a - \alpha c - d - dx_1^2) + (b - \alpha d)x_2^2 - cx_1^2 - cx_1^4$$

which is clearly negative definite if b, c and d are constants with

$$\alpha d > b, \quad c > 0 \quad \text{and} \quad a = \alpha c + d + dx_1^2 .$$

Using these parameters in eq. (3.14) gives $b = c$, so that we require

$$\alpha d > b = c > 0.$$

From eq. (3.12)

$$V(x) = \int_0^{x_1} ax_1dx_1 + \int_0^{x_2} (cx_1 + dx_2)\ dx_2$$

and on substituting for a and integrating

$$V(x) = (\alpha cx_1^2/2) + (dx_1^2/2) + (dx_1^4/4) + cx_1x_2 + dx_2^2/2.$$

This is clearly positive definite if $\alpha c + d > 0$ and $(\alpha c + d)d > c^2$ which will be satisfied by the previous conditions when α and d are positive. Thus V(x) is a Liapunov function for this choice of the parameters and the system is stable for $\alpha > 0$.

Lur'e method

Lur'e considered the stability of a feedback system with a single nonlinearity defined by

$$\dot{x} = Ax + bu$$
$$u = n(\sigma) \qquad (3.15)$$
$$\sigma = c^T x$$

which for a zero input r(t) corresponds to the configuration of Fig. 1.1, although the block diagram to represent eqs. (3.15) is often drawn with the nonlinearity $n(\sigma)$ in the feedback loop. His choice of V function was a quadratic form plus integral of the nonlinearity, that is

$$V(x) = x^T P x + \beta \int_0^{\sigma} n(\sigma)\, d\sigma \qquad (3.16)$$

where β is a constant.

Choosing, for the previous example

$$V(x) = x_2^2 + \beta \int_0^{\sigma} (x_1 + x_1^3)\, dx ,$$

gives

$$\dot{V}(x) = 2x_2\dot{x}_2 + \beta(x_1 + x_1^3)\, \dot{x}_1$$

and on substituting for $\dot{x}_1$ and $\dot{x}_2$ from eqs. (3.13) yields

$$\dot{V}(x) = -2\alpha x_2^2 .$$

This function is negative definite at all points in the x_1–x_2 plane apart from the origin and the line $x_2=0$ if $\alpha>0$. It thus satisfies the conditions of theorem 3 for asymptotic stability.

3.3 FREQUENCY DOMAIN RESULTS

Several additional frequency domain results have been obtained since Popov's original contribution. Some have simple graphical interpretations, an aspect which is considered in the next section. All the criteria can be expressed in terms of the requirement that a specific frequency function, $H(s)$, be positive real. Before presenting the criteria several definitions are required.

Definition 1

A rational function $H(s)$ with real coefficients belongs to the class of positive real functions, that is $H(s) \in \{PR\}$ if:

(a) It is analytic in the open right half s plane

(b) The mapping of $H(s)$ along Γ in the s plane lies in the half plane $\mathrm{Re}\, H \geq 0$, where Γ is the standard

Nyquist contour along the imaginary axis and closed by the infinite semicircle in the right hand side s plane.

Definition 2

A rational function $H(s)$ is strictly positive real, that is $H(s)\in\{SPR\}$, if $H(s-\varepsilon)\in\{PR\}$ for some real $\varepsilon>0$.

Definition 3

The A matrix of a transfer function $G(s)$ is Hurwitz if all of its eigenvalues lie in the left hand side of the s plane. This will be denoted by $A\in\{A_1\}$.

Definition 4

If the A matrix is Hurwitz apart from one eigenvalue at $s=0$ then $A\in\{A_o\}$.

Definition 5

If the closed loop feedback system of Fig. 1.1 has only eigenvalues in the left hand side s plane, when the nonlinearity is replaced by a linear gain k, this will be denoted by $A_k\in\{A_1\}$.

Definition 6 - Sector bounded nonlinearity

A nonlinearity $n(x)$ which is a real, continuous, single valued, scalar function of x, with $n(0)=0$, belongs to the class $\{N\}$ if $k_1\leq n(x)/x\leq k_2$, for $x\neq 0$ and is denoted by $n(x)\in[k_1,k_2]$.

Definition 7 - Slope bounded nonlinearity

A nonlinearity $n(x)$ which is a real, continuous, single valued, scalar function of x, with $n(0)=0$, belongs to the class $\{M\}$ if $m_1\leq dn(x)/dx\leq m_2$, which is denoted by $n'(x)\in[m_1,m_2]$.

Definition 8 - Monotonic nonlinearity

A nonlinearity satisfying $n(x)\in M$ and also $m_1\leq[n(x_1)-n(x_2)]/(x_1-x_2)\leq m_2$ for all finite x_1 and $x_2\neq x_1$ belongs the class $\{M_m\}$.

Definition 9 - Monotonic odd nonlinearity

A nonlinearity satisfying $n(x)=-n(-x)$ and $n(x)\in\{M_m\}$ belongs the class $\{M_{mo}\}$.

The pole and zero transformations which do not effect the stability of the system of Fig. 1.1 with $r(t)=0$ are particularly useful and are given below.

3.3.1 The pole transformation

The pole transformation is illustrated in Fig. 3.1. When the system is transformed by this technique the equivalent linear system $G_\rho(s)$ is given by

$$G_\rho(s) = G(s)/\{1 + \rho G(s)\} \tag{3.17}$$

and if $n(x)\in[k_1,k_2]$ then $n_\rho(x)\in[k_1-\rho,\ k_2-\rho]$.
From eq. (3.17) it is seen that the transformation shifts the poles of G(s), hence the name pole transformation. It is also of value to note that the root locus plots of the transfer functions G(s) and $G_\rho(s)$ for gains from $-\infty$ to ∞ are identical in shape, the only difference being the gain values on the plots. If K is the gain on the plot for G(s) and K_ρ that on the plot for $G_\rho(s)$ then

$$K_\rho = K - \rho \tag{3.18}$$

The transformation alters the nonlinearity sector but keeps the sector width, defined by $k_1 - k_2$ for $n(x)\in[k_1,k_2]$, constant. Similarly, if $n'(x)\in[m_1,m_2]$ then $n'_\rho(x)\in[m_1-\rho,m_2-\rho]$ and the slope sector width is unchanged. In particular if $n(x)\in[0,k_2-k_1]$, and ρ is chosen equal to k_1 then $n_\rho(x)\in[0,k_2-k_1]$, and if $n'(x)\in\{M\}$ then one can choose ρ so that $n'_\rho(x)\in\{M_m\}$.

3.3.2 The zero transformation

In this case the added paths of gain σ are opposite to those of the pole transformation, as shown in Fig. 3.2. Here

$$G_\sigma(s) = G(s) + \sigma \tag{3.19}$$

so that the transformation shifts the zeros of the

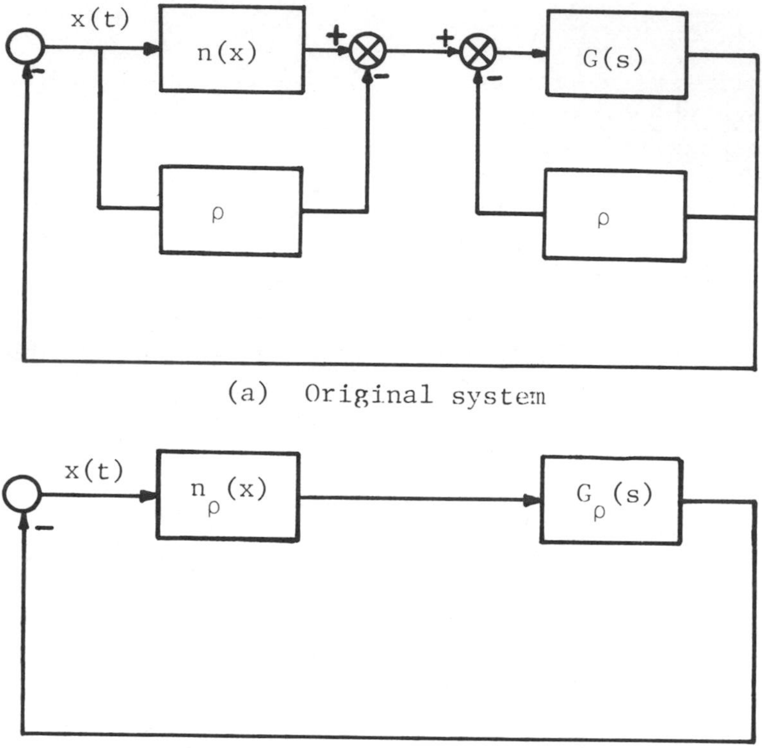

(a) Original system

(b) Pole transformed

Figure 3.1 The pole transformation

linear plant but leaves the poles unaltered. Again, the root locus plots of $G_\sigma(s)$ and G(s) are identical in shape but with the gain markings related by

$$K_\sigma = K/(1 - \sigma K). \tag{3.20}$$

For the nonlinearity it is easily shown that if

$$n(x) \in [k_1, k_2] \text{ then } n_\sigma(x) \in [k_1/(1-\sigma k_1),\ k_2/(1-\sigma k_2)].$$

Thus this transformation, unlike the pole transformation, alters the sector width. A second major difference is that while single valued nonlinearities remain single valued under a pole transformation they may be changed to double valued nonlinearities, and vice versa, under a zero transformation.

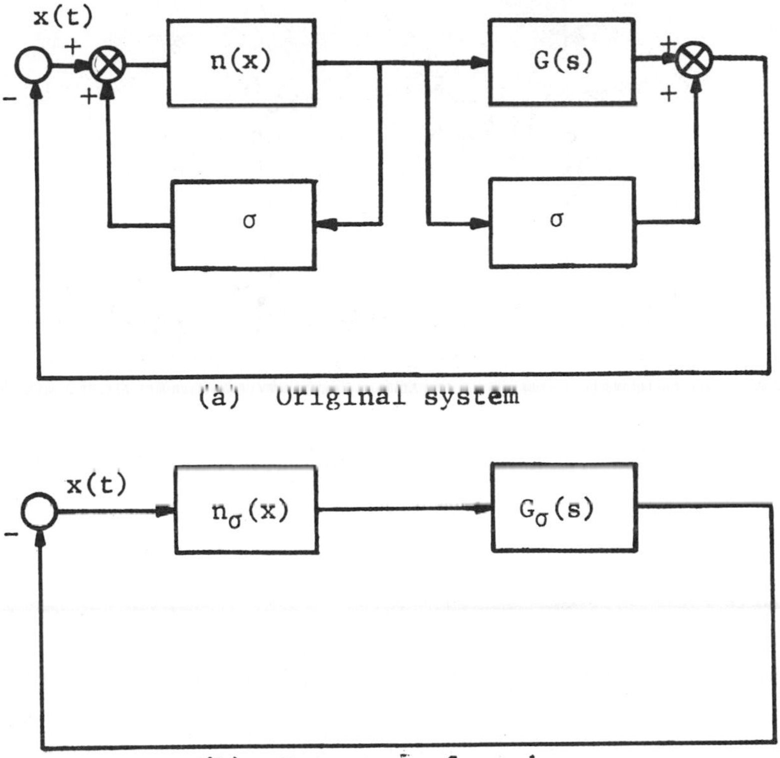

Figure 3.2 The zero transformation

3.3.3 A frequency criterion

We are now in a position to state a general frequency criterion for nonlinear system stability [4].

A nonlinear autonomous system of the form of Fig. 1.1 will be stable if the function $H(s) \in \{PR\}$ where

$$H(s) = \frac{1 + \beta G(s)}{1 + \alpha G(s)} Z(s) \qquad (3.21)$$

and the Z(s) are such that Z(s) or $Z^{-1}(s)$ belong to the class $Z_N(s)$ of frequency domain multipliers the allowable members of which are determined by n(x) as follows: -

(i) If $n(x) \in \{N\}$ then $Z_N(s) = \beta_o s + \gamma_o$ with $\gamma_o > 0$, $\beta_o \geq 0$ and $\alpha = k_1$, $\beta = k_2$ with $A^o_{k_1}$ and $A_{k_2} \in \{A^o_1\}$.

(ii) If $n(x) \in \{M_m\}$ then $Z_N(s)=Z_{RL}(s)$, $\alpha=m_1$, $\beta=m_2$ with A_{m_1} and $A_{m_2} \in \{A_1\}$. $Z_{RL}(s)$ is the driving point impedance of an RL network so that the multiplier Z(s) will consist of alternating, simple, non zero, poles and zeros.

(iii) If $n(x) \in \{M_{mo}\}$ then $Z_N(s)=Z_{RLC}(s)$, $\alpha=m_1$, $\beta=m_2$ with A_{m_1} and $A_{m_2} \in \{A_1\}$. Here $Z_{RLC}(s)$ is a special class of RLC multiplier [4].

It is regularly desirable to make the lower slope or sector limit zero, which is easily done using the pole transformation. For this case the H(s) function is usually written

$$H(s) = (G(s) + \beta^{-1})Z(s) \qquad (3.22)$$

Further for this case we often need to consider transfer functions G(s) which have an integration, that is $A \in \{A_o\}$. This means $A_\alpha \in \{A_o\}$, since $\alpha=0$, and the requirement for stability becomes $H(s) \in \{SPR\}$, where H(s) is given by eq. (3.22).

Use of the above results is greatly facilitated by graphical interpretations and these are considered in the following sections.

3.4 THE POPOV CRITERION

The Popov criterion is obtained from eq. (3.22) using the multiplier $Z(s)=1+qs$ for $q>0$ and $(1+|q|s)^{-1}$ for $q<0$. For example taking $Z(s)=1+qs$ and $n(x) \in \{0,k\}$ eq. (3.22) gives

$$H(j\omega) = U(\omega) + k^{-1} - \omega q\, V(\omega) + j\{V(\omega) + \omega q\, (U(\omega) + k^{-1})\}$$

where $G(j\omega)=U(\omega) + jV(\omega)$.

Re $H(j\omega)$ is an even function of ω and $H(j\omega)$ is positive real if

$$U(\omega) + k^{-1} - \omega q V(\omega) \geq 0, \quad \omega \in (-\infty, \infty). \tag{3.23}$$

Writing $\tilde{X}=U(\omega)$ and $\tilde{Y}=\omega V(\omega)$ and defining

$$\tilde{G}(j\omega) = \tilde{X} + j\tilde{Y} = U(\omega) + j\omega V(\omega) \tag{3.24}$$

eq. (3.23) becomes

$$\tilde{X} + k^{-1} - q\tilde{Y} \geq 0 . \tag{3.25}$$

With the equality sign this is a straight line in the $\tilde{X}$-$\tilde{Y}$ plane of slope q^{-1} passing through the point $(-k^{-1},0)$. The function $\tilde{G}(j\omega)$ is known as the Popov locus and the general result can be summarised as follows.

Theorem 6

A sufficient condition for the autonomous system of Fig. 1.1 with $A_k \in \{A_1\}$ for k=0 and $n(x) \in [0,k]$ to be stable is that the Popov locus $\tilde{G}(j\omega)$ lies to the right of the straight line with non zero slope passing through the point $(-k^{-1},0)$, provided $\lim_{s\to\infty} G(s) > -k^{-1}$. The mathematical condition, eq. (3.23), for stability is usually stated as

$$\mathrm{Re}\{(1 + j\omega q)G(j\omega)\} + k^{-1} \geq \delta > 0 \tag{3.26}$$

where δ is an arbitrary small positive constant. When $A_k \in \{A_o\}$ for k=0 the result is true for $n(x) \in [\varepsilon,k]$ where ε is a small positive constant.

Remark

Using functional analysis it can be shown that eq. (3.26) with $q \geq 0$ is sufficient for L_2 stability if the system of Fig. 1.1 has an input $r(t) \in \{L_2\}$, $n(x) \in [0,k]$ and G(s) is any stable transfer function.

The graphical interpretation of the Popov criterion is shown in Fig. 3.3 for two different $\tilde{G}(j\omega)$ loci. Assuming that these transfer functions are such that the feedback system will be stable for linear gains in the sector $[0,k_H]$, which is known as the Hurwitz sector, then we see that the Hurwitz and Popov sectors for stability are the same in Fig. 3.3(a) but,

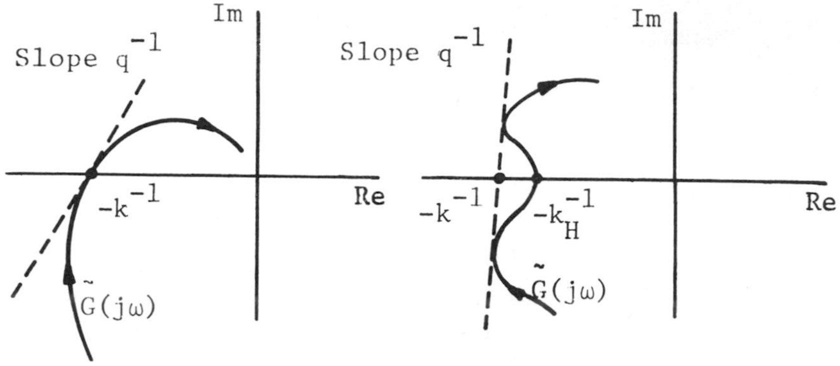

Figure 3.3 Illustration of Popov criterion

since $k<k_H$, the Popov sector is smaller in Fig. 3.3 (b). In order to illustrate certain aspects regarding the application of the Popov criterion we consider several examples.

<u>Example 1</u>

Determine the Popov sector for $G(s)=(1-s)/2(1+s)$. Here $G(j\omega)=(1-j\omega)^2/2(1+\omega^2)$ so that

$$\tilde{G}(j\omega) \;=\; (1 - \omega^2 - 2j\omega^2)/2(1 + \omega^2) = \tilde{X} + j\tilde{Y}.$$

Eliminating ω shows that $\tilde{G}(j\omega)$ is the straight line

$$\tilde{Y} \;=\; \tilde{X} - (1/2)$$

which starts at $(1/2,0)$ for $\omega=0$ and finishes at $(-1/2,-1)$ for $\omega=\infty$. A straight line can thus be drawn through the origin with slope approximately 2 so that the Popov locus remains to its right. This corresponds to $k=\infty$, a value which violates the condition $\lim_{s\to\infty} G(s)>-k^{-1}$ as this yields $k<2$. This latter condition is required to ensure $A_k \in \{A_1\}$. The Popov sector is therefore $(0,2)$ which compares with the Hurwitz sector of $(-2,2)$ for the linear system.

<u>Example 2</u>

A system has a linear transfer function $G(s)=-1/$

$(1-sT_1)(1+sT_2)$ and a nonlinear element $n(x)\in(k_1,k_2)$, find k_1 and k_2 for absolute stability.

For this situation the Popov criterion cannot be applied directly as G(s) is unstable. We therefore use a pole transformation, which from eq. (3.17) gives

$$G_\rho(s) = 1/(s^2T_1T_2 + s(T_1 - T_2) + \rho - 1)$$

and $n_\rho(x)\in(k_1-\rho,\ k_2-\rho)$. For $G_\rho(s)$ to be stable we must have $T_1>T_2$ and $\rho>1$. Further since the locus $G_\rho(j\omega)$ of a second order system is completely contained in the lower half plane so also must be the locus of $G_\rho(j\omega)$. Thus the Popov sector for $n_\rho(x)$ is $(0,\infty)$ provided $T_1>T_2$, which gives a stable sector for n(x) of $(1,\infty)$ in agreement with the Hurwitz sector.

Example 3

A system contains a relay with dead zone and hysteresis, as shown in Fig. 2.5, and a linear transfer function $K/(s+a)^2$. Find conditions for absolute stability.

Here we note that the relay with dead zone and hysteresis can be obtained from the relay with dead zone using positive feedback as shown in Fig. 3.4.

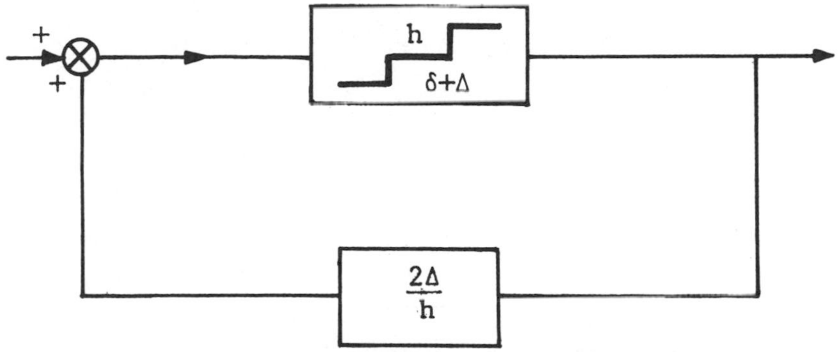

Figure 3.4 Relay with hysteresis

The given feedback system, using a zero transformation, is thus equivalent to a relay with dead zone of width $2(\delta+\Delta)$ and height h and a linear transfer function $G_\sigma(s)=\{K/(s+a)^2\} - 2\Delta/h$. Since $\tilde{G}_\sigma(j\omega)$ lies entirely in the lower half plane the Popov criterion requires $\lim_{s\to\infty} G_\sigma(s) > -k^{-1}$, which gives $2\Delta/h < k^{-1}$. The sector occupied by the relay is $[0, h/(\delta+\Delta)]$ so that for stability

$$2\Delta/h < (\delta+\Delta)/h \text{ , i.e. } \delta > \Delta.$$

3.4.1 Popov sector width

An interesting aspect regarding the pole transformation of section 3.3.1 is that the Popov sector width, unlike the Hurwitz sector width, varies for different values of ρ in the pole transformation. Table 3.1 illustrates this point for a transfer function

$$G(s) = (s + 1)/s(s^3 + 0.6s^2 + 9.5s + 0.9)$$

TABLE 3.1

Variation of sector width with a pole transformation

ρ	Hurwitz Sector	Popov Sector
0	(0, 4.55)	(ε, 0.37)
0.1	(-0.1, 4.55)	(0, 0.94)
0.5	(-0.5, 4.05)	(0, 2.08)
1.2	(-1.2, 3.35)	(0, 3.30)
1.5	(-1.5, 3.05)	(0, 3.00)

A feedback system with $n(x)$ and $G(s)$ will thus be stable for $n(x)\in(\varepsilon, 0.37)$ or $n(x)\in(1.2, 4.5)$, the latter corresponding to an appreciably larger sector width.

3.5 CIRCLE CRITERIA

There are two well known circle criteria both of which have graphical interpretations involving the

Nyquist locus $G(j\omega)$ of the linear elements. One criterion, usually known as the circle criterion [11], is for $n(x)\in\{N\}$ and the other, the off axis circle criterion [12] is for $n(x)\in\{M_m\}$.

The circle criterion can be derived for some categories of $G(s)$ using the multiplier $Z(s)=1$ in eq. (3.21), which is of course equivalent to $q=0$ in the Popov criterion. More recent results have extended the criterion to include transfer functions, $G(s)$, with right hand side s plane poles [5]. Defining the circle or disk $D(k_1,k_2)$ as the circle with its centre on the real axis of the Nyquist diagram and passing through the points $(-k_1^{-1},0)$ and $(-k_2^{-1},0)$, the general form of the circle criterion can be stated as follows: -

Theorem 7

The autonomous system of Fig. 1.1 will be stable for $n(x)\in\{N\}$ and a linear transfer function $G(s)$ if

(i) for $0<k_1<k_2$ the Nyquist plot of $G(j\omega)$ encircles the disk $D(k_1,k_2)$ P times counter-clockwise, where P is the number of poles of $G(s)$ with positive real parts,

(ii) for $0=k_1<k_2$ the Nyquist plot of $G(j\omega)$ satisfies $\text{Re}\ G(j\omega)>-k_2^{-1}$ and $P=0$,

(iii) for $k_1<0<k_2$ the Nyquist plot of $G(j\omega)$ lies in the interior of the disc $D(k_1,k_2)$.

Remark

These criteria are also sufficient for L_2 stability if $r(t)\in\{L_2\}$.

This criterion, which corresponds to the Nyquist criterion for linear systems but with the (-1,0) point becoming a circle, is easy to apply and, using the Nyquist locus, is more amenable for use with frequency domain design techniques and easier to compare with methods such as the DF. The major disadvantage is that being a special case of the Popov criterion it produces more conservative results.

When $n(x)\in\{M_m\}$ the multiplier $Z(s)$ which may be

used in eq. (3.21), $Z(s)^{\pm 1} \in Z_{RL}(s)$. Cho and Narendra [12] have shown that the off axis circle criterion result provides a sufficient condition to be satisfied by the Nyquist locus of G(s) that ensures the existence of such a multiplier for satisfaction of eq. (3.21). The criterion involves the disc $D'(m_1,m_2)$, a circle which passes through the points $(-m_1^{-1},0)$ and $(-m_2^{-1},0)$ and with centre anywhere in the Nyquist plane.

Theorem 8

The autonomous system of Fig. 1.1 with G(s) such that A_{m_1} and $A_{m_2} \in \{A_1\}$ and $n(x)\in\{M_m\}$ with $n'(x)\in[m_1,m_2]$ is stable if the Nyquist locus $G(j\omega)$ for $\omega\in[0,\infty]$ does not enter the disc $D'(m_1,m_2)$.

Remark

For the special case $m_1=0$ the disc becomes a straight line and the requirement is that for $n(x)\in(0,m_2)$ with $A_{m_1}\in\{A_o\}$ or $n(x)\in[0,m_2]$ with $A_{m_1}\in\{A_1\}$ the Nyquist locus of G(s) for $\omega\in[0,\infty]$ lies to the right of a straight line with non zero slope passing through the point $(-m_2^{-1},0)$.

Of the three criteria discussed above the off axis circle criterion usually gives the best results as illustrated by the following example.

Example

Find the conditions for absolute stability of a system with $n(x)\in\{N\}$ and $\{M\}$, $n'(x)\in(0,k)$ and a transfer function $G(s)=(s+0.8)/\{s(s+0.1)(s^2+0.3s+9)\}$.

Fig. 3.5 shows the graph of $G(j\omega)$ and $\tilde{G}(j\omega)$. As $\omega\to 0$, Re $G(j\omega)\to -7.81$ so that the circle criterion requires $k\leq 0.128$ for stability. The Popov line cuts the negative real axis at approximately -3.4 requiring $k \leq 0.29$ to satisfy the Popov criterion, and for the straight line of the off axis circle criterion the corresponding values are -0.48 and $k\leq 2.08$. These values compare with the maximum gain for stability of 2.61 for the linear system.

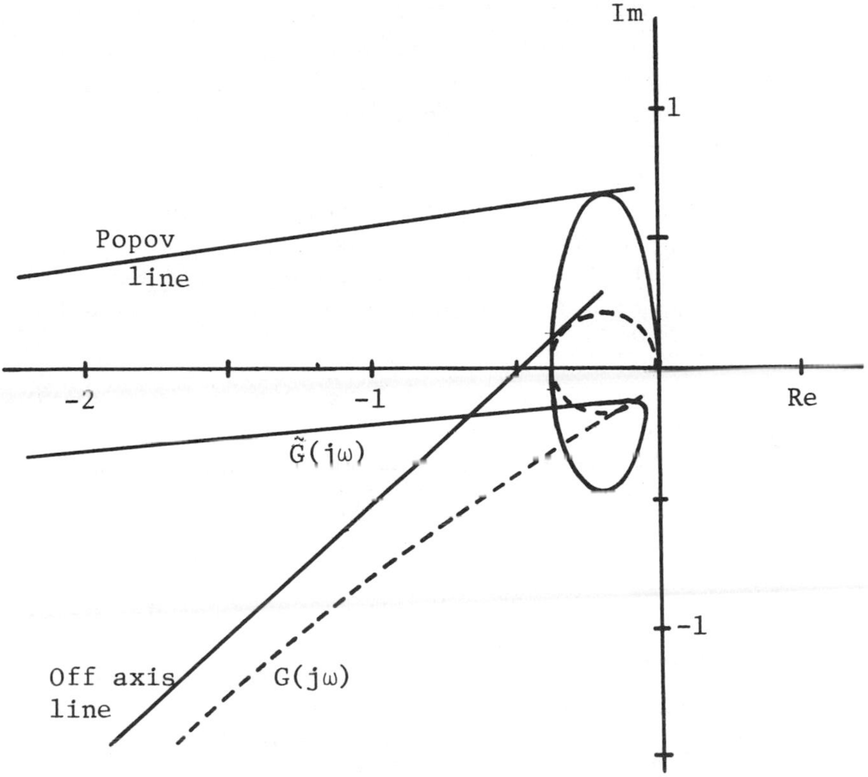

Figure 3.5 Frequency response loci for example

3.6 OTHER CRITERIA

In this section we discuss briefly three other absolute stability criteria. The first, the parabola criterion [13], is a method for applying a less stronger condition than the Popov criterion to obtain conditions for stability when $n(x)\epsilon(k_1,k_2)$, without the need to first apply a pole transformation to reduce the lower bound to zero. The second result due to Dewey and Jury [14], is an extension of the Popov criterion to include both sector and slope bounds of the nonlinearity. Finally an entirely different form of criterion due to Haddad [15] is given. This criterion uses time domain, rather than frequency domain, information on the linear transfer function and a complete description, rather than bounds, of the nonlinearity.

3.6.1 Parabola criterion

With the Popov criterion it is not easy to determine the maximum sector width for stability when $n(x) \in (k_1, k_2)$. As we have seen in Table 3.1 the sector width varies with k_1 and to determine the value of k_2 corresponding to k_1 it is necessary to do a pole transformation and then a Popov plot of $\tilde{G}_\rho(j\omega)$ with $\rho = k_1$ for each choice of k_1. Using eq. (3.21) and a multiplier Z(s) of the Popov form Bergen and Sapiro [13] showed that a sufficient condition for $H(s) \in \{PR\}$ provided $k_1 k_2 > 0$ is

$$(k_1\tilde{X} + 1)(k_2\tilde{X} + 1) \geq q(k_2 - k_1)\tilde{Y} \qquad (3.27)$$

Eq. (3.27) is a parabola in the $\tilde{X}$-$\tilde{Y}$ plane.

Theorem 9

For a system with $n(x) \in [k_1, k_2]$ a sufficient condition for stability is that the Popov locus $\tilde{G}(j\omega)$ not intersect the parabola of eq. (3.27) and $A_k \in \{A_1\}$ for $k \in [k_1, k_2]$ when $k_1 k_2 > 0$.

The parabola of eq. (3.27) can be drawn relatively easily and has the following properties: -

(i) It crosses the real axis, $\tilde{Y}=0$, at $\tilde{X}=-k_1^{-1}$ and $-k_2^{-1}$.

(ii) The parabola is tangent to straight lines drawn through these crossing points of slopes $-q^{-1}$ and q^{-1} respectively.

(iii) The vertex of the parabola, that is where $d\tilde{Y}/d\tilde{X}=0$ is at the point $(\tilde{x}, \tilde{y})$ where $\tilde{x}=-0.5(k_1^{-1}+k_2^{-1})$ and $\tilde{y}=-(k_2-k_1)/4qk_1k_2$.

(iv) The two tangents in (ii) intersect at the point $(\tilde{x}, 2\tilde{y})$.

(v) The vertex of the parabola is a minimum for $q>0$ and a maximum for $q<0$.

3.6.2 The Dewey and Jury criterion

This criterion [14], which was proved using

functional analysis techniques, is an extension of the Popov criterion since slope, in addition to sector, information on the nonlinearity is used.

Theorem 10

For the nonlinear system of Fig. 1.1 if the linear elements G(s) have $A \in \{A_1\}$ and the nonlinearity n(x) is such that $n(x) \in [0,k]$ and $n'(x) \in [k_1,k_2]$ then the system will be stable if there exists a finite number q and a finite number $\mu \geq 0$ such that for all $\omega \geq 0$.

$$H(\omega) = \text{Re}(1 + j\omega q)G(j\omega) + k^{-1} + \mu\omega^2\{1 + (k_2 + k_1)\text{Re } G(j\omega) + k_1k_2|G(j\omega)|^2\} > 0 \qquad (3.28)$$

and in addition

$$\lim_{|x|\to\infty} q[\int_0^x n(x)dx - (xn(x)/2)] = +\infty. \qquad (3.29)$$

A difficulty with this method is the determination of suitable values of q and μ, if any exist, to ensure satisfaction of eq. (3.28). Graphical methods are possible and are given in reference [13] for some special values of k_1 and k_2. A general approach given in reference [15] is simply to use the factor dependent on μ to shift the Popov locus. Eq. (3.28) can be written in the form

$$\text{Re }\{(1 + j\omega q)(G(j\omega) + v)\} + k^{-1} > 0 \qquad (3.30)$$

where v is the factor involving μ which can be written

$$v = \mu\omega^2\{[1 + k_1G^*(j\omega)][1 + k_2G(j\omega)]\} \qquad (3.31)$$

where * denotes the complex conjugate. Eq. (3.30) thus shows that each point on the Popov locus can be shifted to the right by the frequency dependent factor v and the Popov graphical criterion applied to the shifted locus. The shifted locus has, of course, to be plotted for several values of μ to try and find a suitable value. Better results than the

Popov criterion have been obtained for some examples in reference [14] using the above theorem, however, it is not clear whether superior results to those of the off axis circle criterion can be obtained.

3.6.3 The Haddad criterion

The results obtained by Haddad [16] are of interest both because they use an entirely different approach from those considered previously and in doing so include considerably more information about the nonlinearity. Here, we restrict our attention to the presentation of his results for the autonomous system of Fig. 1.1 with a nonlinearity in the first and third quadrants only. Additional theorems are given in reference [16] to cover the situations of bounded inputs and the nonlinearity existing in any quadrant.

Two positive numbers, A_+ and A_- are used to represent the linear transfer function G(s) in this criterion. A_+ is the area enclosed by the impulse response g(t) of G(s) above the time axis and A_- that enclosed below. A first and third quadrant nonlinearity is assumed to be split into two characteristics, $n_+(x_+)$ in the first quadrant and $n_-(x_-)$ in the third quadrant, where $x_+=x$ for $x>0$ and $x_-=-x$ for $x<0$. Two loci $L_1(x_+,x_-)=0$ and $L_2(x_+,x_-)=0$ are then plotted in the x_+-x_- plane where

$$L_1(x_+,x_-) = x_+ - A_-n_+(x_+) - A_+n_-(x_-) \qquad (3.32)$$

and

$$L_2(x_+,x_-) = x_- - A_+n_+(x_+) - A_-n_-(x_-) \qquad (3.33)$$

respectively.

Theorem 11

The system of Fig. 1.1 is stable if $S_1 \cap S_2=0$, where S_1 denotes those points in the x_+-x_- plane satisfying $L_1(x_+,x_-)<0$ and S_2 those satisfying $L_2(x_+,x_-)<0$. When $S_1 \cap S_2 \neq 0$ but is finite the maximum possible amplitudes x_{+m} and x_{-m} at the nonlinearity are the maximum values of x_+ and x_- in $S_1 \cap S_2$.

Fig 3.6 shows a graphical interpretation of this theorem. Since for this sketch $S_1 \cap S_2$ is finite one cannot conclude the system is stable, although this may in fact be the case, but simply that $x_{+m}=b_1$ and $x_{-m}=b_2$. Thus, if a limit cycle exists the input to the nonlinearity will lie between $-b_2$ and b_1.

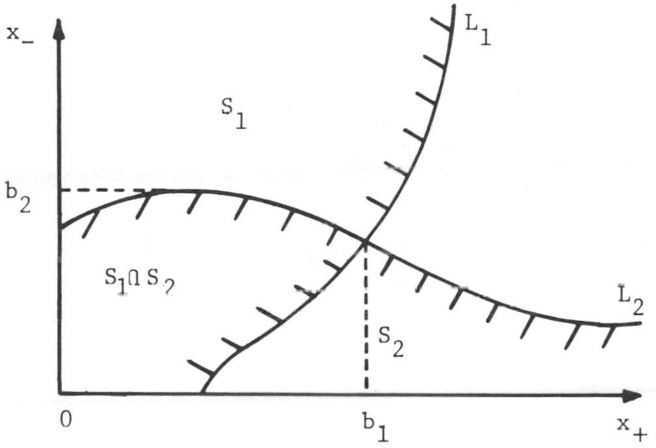

Figure 3.6 Graph to illustrate the Haddad criterion

3.7 LIMIT CYCLE CRITERIA

Aizerman [17] conjectured that if the system of Fig. 1.1 is stable with k replacing n(x) for $k_1<k<k_2$ then the nonlinear system would be stable for $n(x) \in (k_1, k_2)$. This conjecture is not true and is considered in more detail in Chapter 5. It is of further interest to know if the system initial condition response can become unbounded when the above conditions apply. This question was answered to a large extent by Vogt and George [18] who showed that if G(s) is such that $A_k \in \{A_1\}$ for $k_1<k<k_2$ and $(n(x)/x) \to k$ as $x \to \infty$ then all solutions of the autonomous system of Fig. 1.1 are bounded. This means, therefore, that for nonlinearities which approach a straight line within the Hurwitz sector of the system for large inputs, the only possible form of instability is a limit cycle.

In situations where absolute stability cannot be proved several results are available which rule out the possibility of limit cycles with certain frequencies. The original results in this area are due to Garber [19] who proved the following theorem.

Theorem 12

For the system of Fig. 1.1 and the conditions of the Popov criterion no symmetrical odd limit cycle of fundamental frequency ω can exist if

$$\mathrm{Re}\,\{(1 + jm\omega q)G(jm\omega)\} + k^{-1} > 0 \qquad (3.34)$$

for all $m=2n+1$, $n=1,2,\ldots\infty$.

The simple graphical interpretation of this result is that a limit cycle of frequency ω cannot exist if all the frequencies $m\omega$ lie to the right of the Popov line. Garber also showed that three additional terms could be added to the left hand side of the inequality (3.34) if the slope bounds of the nonlinearity are taken into account. One of these additional terms is identical with that added to the Popov criterion to produce the Dewey and Jury criterion.

A valuable alternative result was also obtained by Garber by his functional analysis approach. The theorem which can be stated as follows can be used when G(s) has right hand side s plane poles.

Theorem 13

For the system of Fig. 1.1 with $n(x)\in[0,k]$ and $\hat{G}(s)$ a stable transfer function, where $\hat{G}(s)$ is the inverse of G(s), no limit cycle of frequency ω can exist if

$$\mathrm{Re}\{(1 + jm\omega q)\hat{G}(jm\omega)\} + k < 0 \qquad (3.35)$$

for $m=2n+1$, $n=0,1,2,\ldots\infty$.

This result again has a graphical interpretation that no limit cycle at a fundamental frequency ω can exist if all frequencies $m\omega$ on the Popov plot of $\hat{G}(j\omega)$ lie to the left of the line of slope q^{-1} drawn through the point (-k,0). It is also possible to add additional terms involving the slope bounds of the nonlinearity to the left hand side of the inequality (3.35) [15]. For the particular case of $n'(x)\in[0,k_2]$ a Nyquist diagram graphical interpretation of this criterion, with a single additional term of the Dewey and Jury form added to the inequality, has been given in reference [20].

A further result is due to Cook [21] who has obtained a limit cycle version of the off axis circle criterion. This result shows that for the conditions of the off axis circle criterion no limit cycle can exist at a fundamental frequency ω if all frequencies $m\omega$ lie outside the off axis circle.

The main advantage of all these results is that they provide information on possible limit cycle frequencies which may be of value to other methods for determining if a limit cycle exists and if so at what frequency.

3.8 INSTABILITY CRITERIA

Several results are known which provide sufficient conditions for instability or the existence of a limit cycle in a nonlinear system. These are of some interest in control theory since where instability conditions are satisfied there is no point in trying to prove stability. A Liapunov type condition for instability is provided by the following theorem [22].

Theorem 14

The equilibrium solution of the autonomous system of eq. (3.1) is not asymptotically stable if there exists a scalar function $V(x)$ with the following properties in some closed neighbourhood, R, of the origin: -

(i) $V(x)$ vanishes at the origin, has continuous partial derivatives, and assumes negative values arbitrarily close to the origin; and

(ii) the derivative $\dot{V}(x)$ along the solutions of eq. (3.1) is negative semi-definite.

If $V(x)$ and $\mathring{V}(x)$ are negative definite in R then the null solution is completely unstable, that is any solution will ultimately leave R within a finite time.

Frequency domain results for instability have been obtained by Brockett and Lee [23] which are restrictive in that they require at least one pole of G(s) to lie in the right hand side s plane. The following circle and Popov type criteria were proved using Liapunov type results similar to the above theorem.

Circle Theorem

The autonomous system of Fig. 1.1 with $n(x) \in (k_1, k_2)$ is unstable if the Nyquist locus of G(s) does not intersect and encircles the disk $D(k_1, k_2)$ fewer than P times in the counterclockwise direction, where P is the number of right hand side s plane poles of G(s).

Popov Type Theorem

The autonomous system of Fig. 1.1 with $n(x) \in [0,k]$ and G(s) having at least one right hand side s plane pole is unstable if the graphical Popov condition for stability is satisfied.

More recently several results have been obtained by Noldus [24-26] on conditions for the existence of limit cycles in the system of Fig. 1.1. These criteria involve satisfaction of several conditions including tests on frequency functions of the form of H(s) in eq. (3.21). Currently there is also interest in the use of the Hopf bifurcation theorem for determining the existence of limit cycles in feedback systems [27,28]. Loosely, the theorem states that if on linearizing the system equation about an equilibrium point a pair of complex conjugate eigenvalues cross the imaginary axis as a parameter μ varies through certain critical values, then for near-critical values of μ there are limit cycles close to the equilibrium point. The interested reader is referred to the relevant papers for further details of these methods.

3.9 SUMMARY

Various results have been given in this chapter which can be used for investigating the stability of the system of Fig. 1.1. The criteria presented only provide sufficient conditions for stability and in some instances they produce very conservative results primarily because they allow for a large class of nonlinear characteristics. The frequency domain criteria having graphical interpretations are easily used and the incorporation of one or more additional parameters, as in the Dewey and Jury criterion, is not difficult with computer graphics facilities. Although we have restricted our considerations here to an autonomous system, many of the criteria are valid for certain classes of inputs and/or time varying nonlinear characteristics. The reader who is interested in these more general results and also detailed proofs of the various theorems will find them in those references [4-7, 22] which adopt a rigorous theoretical approach.

REFERENCES:

1. Liapunov, A.M.: "Stability in nonlinear control systems", Princeton University Press, Princeton, N.J., 1961.

2. Popov, V.M.: "Absolute stability of nonlinear control systems of automatic control", Automat. and Remote Control, Vol. 22, pp. 857-858, 1962.

3. Aizerman, M.A. and Gantmacher, F.R.: "Absolute stability of regulator systems", Holden Day, San Francisco, 1974.

4. Taylor, J.H. and Narendra, K.S.: "Frequency domain criteria for absolute stability", Academic Press, New York, 1973.

5. Vidyasagar, M.: "Nonlinear systems analysis", Prentice-Hall, N.J., 1978.

6. Desoer, C.A. and Vidyasagar, M.: "Feedback systems: input-output properties", Academic Press, New York, 1975.

7. Holtzman, J.M.: "Nonlinear system theory - a functional analysis approach", Prentice-Hall, Englewood Cliffs, N.J., 1970.

8. Lur'e, A.I.: "Some nonlinear problems in the theory of automatic control", H.M. Stationery Office, London, 1957.

9. LaSalle, J.P. and Lefschetz, S., "Stability by Liapunov's direct method with applications", Academic Press, New York, 1961.

10. Schultz, D.G. and Gibson, J.E.: "The variable gradient method for generating Liapunov functions" Trans. A.I.E.E., 81(II), pp. 203-210, 1962.

11. Sandberg, I. W.: "A frequency domain condition for the stability of feedback systems containing a single time-varying nonlinear element", Bell

System Tech. J., Vol. 43, No. 4, 1964.

12. Cho, Y.S. and Narendra, K.S.: "An off-axis circle criterion for the stability of feedback systems with a monotonic nonlinearity", IEEE Trans. Automat. Contr., Vol. AC-13, No. 4, 1968.

13. Bergen, A.R. and Sapiro, M.A.: "The parabola test for absolute stability", IEEE Trans. Automat. Contr., Vol. AC-12, No. 3, 1967.

14. Dewey, A.G. and Jury, E.I.: "A stability inequality for a class of nonlinear feedback systems", IEEE, Trans. Automat. Contro., Vol. AC-11, No. 1, pp. 54 62, Jan. 1966.

15. El-Sakkary, A.K.: "Stability of nonlinear multivariable control systems", M.Sc. thesis, University of New Brunswick, 1978.

16. Haddad, E.K.: "New criteria for bounded-input bounded-output and asymptotic stability of nonlinear systems", Proc. IFAC 5th World Congress, Paris, Pt. IV, Paper 32.2, 1972.

17. Aizerman, M.A.: "On a problem relating to the global stability of dynamic systems", Uspehi Mat. Nauk, Vol. 4, No. 4, 1949.

18. Vogt, W.G. and George, J.H.: "On Aizerman's conjecture and boundedness", IEEE Trans. Automat. Contr., Vol. AC-12, pp. 338-339, June 1967.

19. Garber, E.D.: "Frequency criteria for the absence of periodic modes:, Automat. and Remote Control, Vol. 28, No. 11, pp. 1776-1780, 1967.

20. Rootenberg, J. and Walk, R.: "Geometric criterion for the absence of limit cycles in nonlinear control systems", IEEE Trans. Automat. Control., Vol. AC-18, pp. 64-65, Feb. 1973.

21. Cook, P.A.: "Conditions for the absence of limit cycles", IEEE Trans. Automat. Contr., Vol. AC-21, pp. 339-345, June, 1976.

22. Willems, J.L.: "Stability theory of dynamical systems", Nelson, London, 1970.

23. Brockett, R.W. and Lee, H.B.: "Frequency domain instability criteria for time-varying and non-linear systems", Proc. IEEE, Vol. 55, No. 5, pp. 604-619, 1967.

24. Noldus, E.J.: "A counterpart of Popov's theorem for the existence of periodic solutions", Int. J. Control, Vol. 13, pp. 705-719, 1971.

25. Noldus, E.J.: "Autonomous periodic motion in nonlinear feedback systems", IEEE Trans. Automat. Control., Vol. AC-19, pp. 381-387, 1974.

26. Noldus, E.J.: "Oscillation criteria of the Popov type", IEEE Trans. Automat. Contr., Vol. AC-20, pp. 577-579, 1975.

27. Mees, A.I. and Chua, L.O.: "The Hopf bifurcation theorem and its applications to nonlinear oscillations in circuits and systems", Electronics Research Laboratory, Report No. M77/63, University of California, Berkeley, 1977.

28. D.J. Allwright: "Harmonic balance and the Hopf bifurcation theorem", Math. Proc. Cambridge Philosophical Society, vol. 82, 1977, pp. 453-467.

PROBLEMS

1. The system of Fig. 1.1 has $n(x) \in [0,k]$ and $G(s) = (s+1)^2/(s^3 + 9s^2 + 26s + 99)$. Determine the maximum value of k for which the system will be stable using the Popov criterion. It is desired that the system be stable for $k=\infty$ and a series compensator $G_c(s) = (1 + 0.25s)/(1 + 0.25\alpha s)$ is added. Find the required value of α to satisfy the Popov criterion.

2. For question 1 find the maximum value of k for stability using the circle criterion. What is the maximum value of k allowed by the circle criterion for stability when the compensator $G_c(s)$ is added with the computed value of α.

3. Repeat question 1 using the off axis circle criterion if the nonlinearity is also monotonic with $n'(x) \in [0,k]$.

4. The system of Fig. 1.1 has $n(x) \in (0,k)$ and $G(s) = (s+1)/\{s(s+0.1)(s^2+0.4s+16)\}$. Determine the maximum value of k, say k_c, for stability using the Popov criterion and the frequency, ω_c, at which the Popov locus cuts the real axis. If $n(x) \in (0,1.1\ k_c)$ would you expect the frequency of any possible limit cycle to be less than or greater than ω_c?

5. Determine for the system of Fig. 1.1 the largest value of k for stability by the Popov and circle criteria, if $n(x) \in [0,k]$ and $G(s)=(s+1)e^{-s}/(s^2+5s+6)$. What is the value of k for the linear system?

6. If $n(x) \in [k_1,k_2]$ and $G(s)=\{(s+1)/\{s(s+0.2)(s^2+0.5s+10)\}^2$ determine the largest sector width for stability using (a) the parabola and (b) the Popov criteria.

7. Prove the parabola criterion from eq. (3.21).

8. What can you conclude about the stability of the linear system $n(x) = kx$ if (a) $G(s)=1/(s+1)(s+2)(s+3)$ and (b) $G(s) = 1/s(s+1)(s+2)$ using the Haddad criterion?

9. A nonlinearity $n(x)$ is defined by

$$n(x) = \begin{cases} k_1 x & x > 0 \\ k_2 x & x < 0 \end{cases}$$

Determine conditions on $k_1 k_2$ for stability by the Haddad criterion if $G(s) = 1/(s+1)(s+2)(s+3)$. How does the maximum value of $k_1 k_2$ for stability compare with (a) the Popov criterion and (b) the Routh criterion (assuming $k_1 = k_2$). Can you prove stability by the Haddad criterion and not by the Popov criterion for particular choices of k_1 and k_2?

10. In the light of the results obtained in questions 8 and 9 discuss the type of systems for which the Haddad criteria may be of more value than the frequency domain methods.

11. Can you obtain better results for problem 8 by first using a pole transformation?

CHAPTER 4

The Describing Function Method

4.1 INTRODUCTION

The DF method is an approximate procedure for investigating the existence of limit cycles in the feedback system of Fig. 1.1. The concept is one of quasilinearization whereby a static nonlinear characteristic is represented by a gain dependent upon the magnitude of the input signal. This gain is evaluated on the assumption that the input signal is a sinusoid. It is found to be a very satisfactory assumption for many feedback control systems due to the low pass filtering action normally provided by the plant transfer function, G(s). For example, if the nonlinearity is an ideal relay, so that its output is an odd symmetric square wave when a limit cycle exists, and if $|G(j\omega)|$ varies as $1/\omega^2$ for all frequencies above the square wave fundamental frequency, the relative magnitude of the harmonics in the output, c(t), of G(s) will be proportional to $1/n^3$, for n=3, 5, etc. The rms value of the output fundamental is within 5% of the total rms output. Because the majority of instabilities in nonlinear systems occur as limit cycles and any other input signal to the nonlinearity may be expected to undergo a gain change similar to that for a sinusoid, the DF method is also used as a stability criterion.

Initial investigations of nonlinear systems using the DF took place in several countries at about the same time [1-5]. The power of the method is its simplicity and the ease with which it can be used with the methods of classical control theory both for analysis and design purposes. Its weakness is that, being an approximate procedure, it may give incorrect results. Situations where the method may fail or give unsatisfactory results are usually detectable if additional calculations are undertaken. The question of DF accuracy will therefore be discussed in some detail in the next chapter.

4.2 EXPRESSIONS FOR DF EVALUATION

As mentioned in the introduction the DF which is also called the sinusoidal describing function since the same concept can be used for other forms of nonlinearity input signal, is the gain of a nonlinearity to a sinusoidal input signal. The nonlinearity output measure used to evaluate the gain is the fundamental component of the output waveform which is, of course, at the same frequency as the input sinusoid. For double valued or frequency dependent nonlinearities the fundamental of the output may not be in phase with the input so that the gain, or DF, is complex. Here we will consider only frequency independent nonlinearities which may be either single or double valued. In addition, we will assume the nonlinear characteristics to be odd symmetrical since, when this is not the case, it is usually necessary to consider a bias input as well as the sinusoid; a topic which is discussed in Section 4.7.

4.2.1 Single valued nonlinearity

For the single valued nonlinearity, SVNL, the DF will be real, that is the output fundamental will be in phase with the input sinusoid. The DF, which we will denote by $N(a)$, where a is the amplitude of the input sinusoid, can be evaluated from a Fourier analysis of the output waveform, $y(\theta)$. This waveform is easily sketched, as illustrated in Fig. 4.1, for an ideal saturation characteristic. For reasons which will become apparent later, when we also consider the harmonic content of $y(\theta)$, we have chosen the origin of the input $x(\theta)$ so that $x(\theta) = a \cos \theta$. The fundamental, a_1 of $y(\theta)$ is then given by

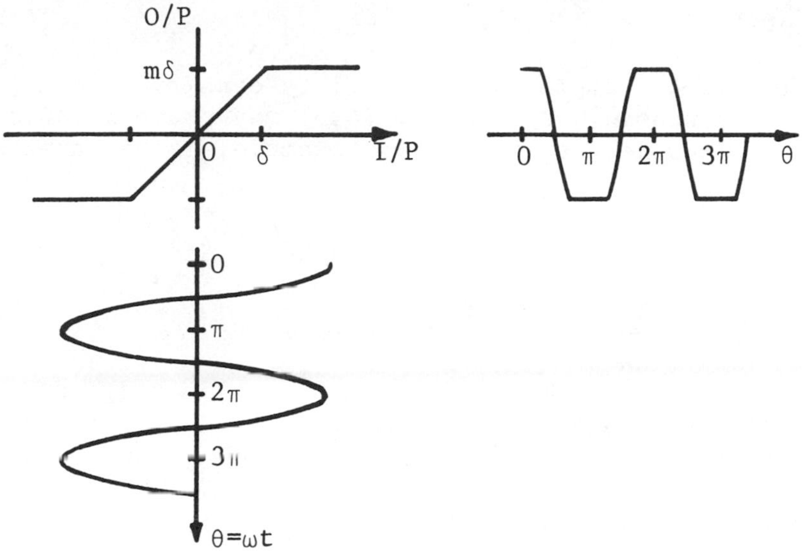

Figure 4.1 Saturation characteristic and waveforms

$$a_1 = (1/\pi) \int_0^{2\pi} y(\theta) \cos \theta \, d\theta \qquad (4.1)$$

which, because the nonlinearity is assumed to be odd symmetric, may be written

$$a_1 = (4/\pi) \int_0^{\pi/2} y(\theta) \cos \theta \, d\theta \qquad (4.2)$$

and the DF, $N(a)$, is given by

$$N(a) = a_1/a = (4/a\pi) \int_0^{\pi/2} y(\theta) \cos \theta \, d\theta. \qquad (4.3)$$

If the equation of the nonlinear characteristic is

$$y = n(x) \qquad (4.4)$$

then substituting for θ in terms of x in eq. (4.3) which since $\cos \theta = x/a$, gives

$$N(a) = (4/a^2) \int_0^{a} xn(x)p(x) \, dx \qquad (4.5)$$

where

$$p(x) = \{\pi(a^2 - x^2)^{1/2}\}^{-1} \tag{4.6}$$

which is the amplitude probability density function for a sinusoid. Use of eq. (4.5) may prove advantageous in some situations, for example, if we integrate by parts

$$N(a) = (4/a\pi)n(0^+) + (4/a^2\pi) \int_0^a n'(x) (a^2 - x^2)^{1/2}\, dx \tag{4.7}$$

where $n'(x) = d\{n(x)\}/dx$ and $n(0^+) = \lim n(\varepsilon)$, $\varepsilon \to 0$, ε positive. $N(a)$ can thus be evaluated from the slope of $n(x)$.

4.2.2. Double valued nonlinearity

Fig. 4.2 shows a general odd symmetric double valued nonlinearity, DVNL, which is defined by

$$y = \begin{cases} n_1(x) & \text{for } dx/d\theta > 0 \\ n_2(x) & \text{for } dx/d\theta < 0. \end{cases} \tag{4.8}$$

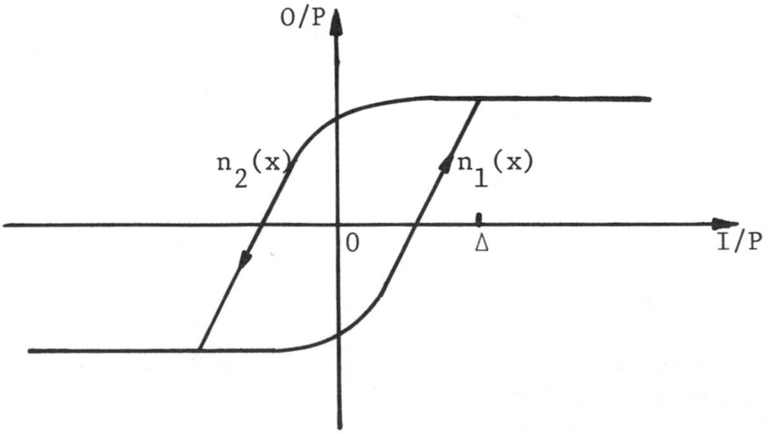

Figure 4.2 Odd symmetrical DVNL

In addition as a consequence of the odd symmetry

$$n_1(x) = -n_2(-x) \tag{4.9}$$

and

$$y(\theta) = -y(-\pi + \theta)\,. \tag{4.10}$$

The magnitudes of the in phase and quadrature components of the output fundamental, assuming $x = a \cos \theta$, are thus given by

$$a_1 = (2/\pi) \int_0^\pi y(\theta) \cos \theta \, d\theta \tag{4.11}$$

and

$$b_1 = (2/\pi) \int_0^\pi y(\theta) \sin \theta \, d\theta \tag{4.12}$$

respectively.

If we define

$$n_p(x) = \{n_1(x) + n_2(x)\}/2 \tag{4.13}$$

and

$$n_q(x) = \{n_2(x) - n_1(x)\}/2 \tag{4.14}$$

where $n_p(x)$ will have odd symmetry and $n_q(x)$ even symmetry, it is easily shown that eqs. (4.11) and (4.12) may be written in terms of the input x as

$$a_1 = (4/a) \int_0^a x n_p(x) p(x) \, dx \tag{4.15}$$

$$b_1 = (4/a\pi) \int_0^a n_q(x) \, dx. \tag{4.16}$$

The DF which is now complex is given by

$$N(a) = (a_1 - jb_1)/a \tag{4.17}$$

which may also be written

$$N(a) = N_p(a) + jN_q(a) \tag{4.18}$$

where $N_p(a) = a_1/a$ and $N_q(a) = -b_1/a$

or in polar coordinates

$$N(a) = M(a)e^{j\psi(a)} \tag{4.19}$$

with $M(a) = (a_1^2 + b_1^2)^{1/2}/a$ and $\psi(a) = -\tan^{-1} b_1/a_1$.

Examination of eq. (4.16) shows that the value of b_1 is proportional to the area of the loop contained within the DVNL. The equation may be written

$$b_1 = (\text{area of DVNL loop})/a\pi. \tag{4.20}$$

A word of caution regarding the use of the above equations involving $n_p(x)$ and $n_q(x)$. The theory assumes that these characteristics are independent of a, that is whatever the input amplitude to the DVNL the same characteristic will be followed. This is not the situation, for example, with hysteresis in a magnetic material where the nonlinear path followed is amplitude dependent. This type of nonlinearity is probably more correctly described as multi-valued rather than double valued and if the above equations are employed the appropriate values of $n_p(x)$ and $n_q(x)$ for the given value of a must be used.

4.3 DFS OF SPECIFIC NONLINEARITIES

To illustrate the application of the above results DFs are derived for some specific nonlinear characteristics. In general since evaluation of the DF expressions for the integrals with respect to x rather than θ avoids at least a sketch of the output waveform y(θ), this approach will be preferred, although both methods have been used in the first example. A detailed listing of DFs for many nonlinear characteristics is given in Table I.1 of Appendix I.

4.3.1 Ideal saturation

An ideal saturation nonlinearity defined by

$$n(x) = \begin{cases} mx & \text{for } -\delta < x < \delta \\ m\delta & \text{for } x \geq \delta \\ -m\delta & \text{for } x \leq -\delta \end{cases} \tag{4.21}$$

is shown in Fig. 4.1 with corresponding input and output waveforms.

Using eq. (4.3)

$$N(a) = (4/a\pi)\{\int_0^\beta m\delta \cos\theta \, d\theta + \int_\beta^{\pi/2} ma \cos^2\theta \, d\theta\} \tag{4.22}$$

where $\beta = \cos^{-1}(\delta/a)$. Integrating gives

$$N(a) = (4m/a\pi)\{\delta \sin\beta + (a/4)(\pi - 2\beta - \sin 2\beta)\}$$

which gives

$$N(a) = (m/\pi)(\sin 2\beta + \pi - 2\beta). \tag{4.23}$$

Alternatively, from eq. (4.7) which uses the slope of the nonlinearity

$$N(a) = (4/a^2\pi) \int_0^\delta m(a^2 - x^2)^{1/2} \, dx \tag{4.24}$$

which gives

$$N(a) = (m/\pi)(2\alpha + \sin 2\alpha) \tag{4.25}$$

where $\alpha = \sin^{-1} \delta/a = (\pi/2) - \beta$, which agrees with equation (4.23). Eq. (4.25) may also be written

$$N(a) = (2m/\pi)\{\sin^{-1}\rho + \rho(1 - \rho^2)^{1/2}\} \tag{4.26}$$

where $\rho = \delta/a$.

The above calculated values for $N(a)$ only apply for $a>\delta$ as $N(a)=m$ for $a<\delta$. The complete expression for $N(a)$ can be written

$$N(a) = mf_1(\rho) \tag{4.27}$$

where the function $f_1(\rho)$ is defined by

$$f_1(\rho) = \begin{cases} 1 & \text{for } \rho>1 \\ (2/\pi)\{\sin^{-1}\rho + \rho(1-\rho^2)\}^{1/2} & \text{for } |\rho|<1 \\ -1 & \text{for } \rho<-1 . \end{cases} \tag{4.28}$$

Since many linear segmented characteristics can be built up from saturation characteristics in parallel, with different values of m and δ, the reasons for defining the function $f_1(\rho)$ are obvious and it occurs regularly in Table I.1.

4.3.2 Odd power law

For the characteristic

$$y = x^n , \text{ n odd} \tag{4.29}$$

application of eq. (4.5) yields

$$N(a) = (4/a^2) \int_0^a x^{n+1} p(x)\, dx$$

$$= 2\mu_{n+1}/a^2 \tag{4.30}$$

where μ_n is the nth moment of the sinusoidal probability density function and is given by

$$\mu_{n+1} = \frac{n(n-2)(n-4)\ldots 3.1}{(n+1)(n-1)(n-5)\ldots 4.2}\, a^{n+1} . \tag{4.31}$$

4.3.3 Relay with hysteresis and dead zone

The characteristic and the corresponding in phase, $n_p(x)$, and quadrature, $n_q(x)$, nonlinearities, which are valid for $a>\delta+\Delta$, are shown in Fig. 4.3. Using eq. (4.15).

$$N_p(a) = (4/a^2)\{ \int_{\delta-\Delta}^{\delta+\Delta} (h/2)xp(x)\, dx$$

$$+ \int_{\delta+\Delta}^{a} hxp(x)\, dx\}.$$

Also

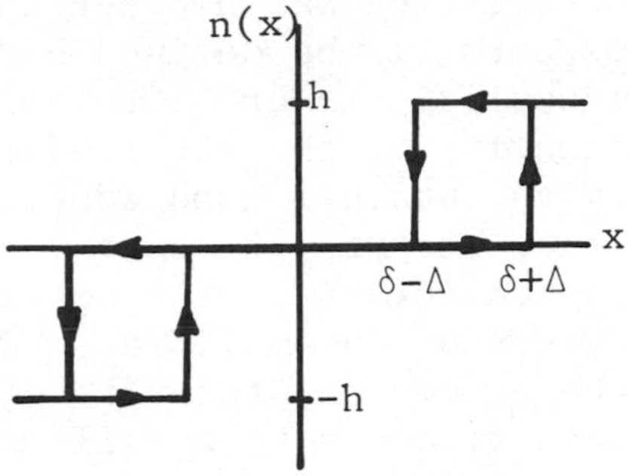

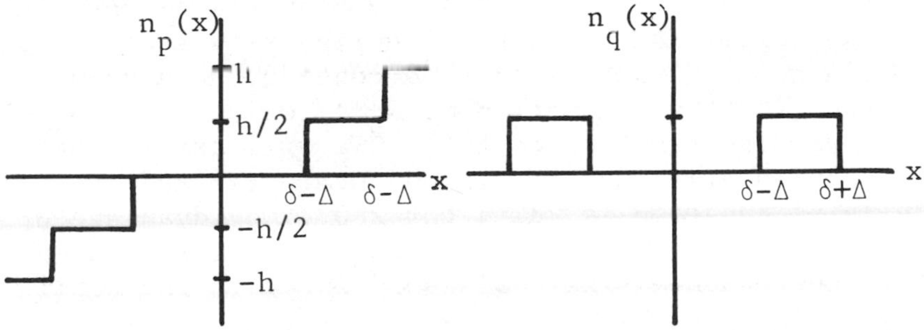

Figure 4.3 Relay with dead zone and hysteresis

$$\int xp(x)\,dx = \int \frac{x\,dx}{\pi(a^2 - x^2)^{1/2}} = -(a^2 - x^2)^{1/2}/\pi \qquad (4.33)$$

so that

$$N_p(a) = (2h/a^2\pi)[\{a^2 - (\delta + \Delta)^2\}^{1/2} + \{a^2 - (\delta - \Delta)^2\}^{1/2}] \text{ for } a > \delta + \Delta \qquad (4\text{-}34)$$
$$= 0 \quad \text{for } a < \delta + \Delta.$$

The quadrature gain $N_q(a)$ for $a>\delta+\Delta$ can be written down immediately using eq. (4.20) to give

$$N_q(a) = 4h\Delta/a^2\pi. \qquad (4\text{-}35)$$

4.4 APPROXIMATE DF EVALUATION METHODS

Nonlinear characteristics which occur in practical systems will not normally be defined by simple mathematical relationships. In order to calculate the DF of the nonlinearity to the required accuracy it may be necessary to obtain a good approximation to the characteristic; better, for example, than the ideal saturation nonlinearity will approximate the saturation behaviour of an amplifier. On the other hand it should be remembered that the DF method is an approximate procedure and nothing will normally be gained by evaluating the DF to an accuracy better than 1%. Several analytical techniques are possible for approximating a measured nonlinearity so that its DF can be easily obtained. We consider here several methods for the evaluation of $N_p(a)$ only, since $N_q(a)$ follows directly from (4.20). We will assume that the nonlinearity, $n_p(x)$, is odd symmetric and the defining equation is valid for x>0. An advantage of all the methods given is that they can also be used for other than a single sinusoid input to the nonlinearity [6].

4.4.1 Polynomial expansion

If the nonlinearity is approximated by the power series

$$n_p(x) = \sum_{m=0}^{M} A_m x^m \tag{4.36}$$

then from eq. (4.15)

$$N_p(a) = (4/a^2) \sum_{m=0}^{M} A_m \int_0^a x^{m+1} p(x)\,dx.$$

The integral can be evaluated in terms of the gamma function to give the result

$$N_p(a) = \sum_{m=0}^{M} 2A_m a^{m-1} \pi^{-1/2} \Gamma\{(m+2)/2\}/ \Gamma\{(m+3)/2\} \quad \text{for } m > -2. \tag{4.37}$$

It is seen from this result that if the polynomial

for x>0 has only odd (even) powers of x then the DF is an even (odd) power series in a.

4.2.2 Quantised approximation

An easy and rapid procedure for approximating a non-linear characteristic, especially if it is defined by a table of coordinates, is to use the quantised approximation illustrated in Fig. 4.4. This can, for example, be done directly in a computer program which uses the table of coordinates as input. Again employing eq. (4.15)

$$N_p(a) = (4/a^2)\{\int_{\delta_1}^{\delta_2} h_1 x p(x)\,dx + \int_{\delta_2}^{\delta_3} (h_1 + h_2) x p(x)\,dx + \ldots \int_{\delta_m}^{a} (h_1 + h_2 \ldots + h_m) x p(x)\,dx\}$$

which on using eq. (4.33) gives

$$N_p(a) = (4/a^2\pi) \sum_{m=1}^{M} h_m (a^2 - \delta_m^2)^{1/2}. \tag{4.38}$$

This result can in fact be written down directly from eq. (4.7) if the integral involving n'(x) is interpreted correctly.

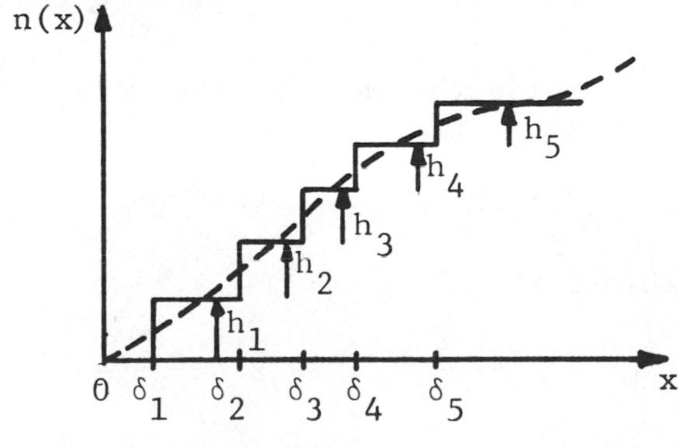

Figure 4.4 A quantised approximation

4.4.3 Fourier series expansion

Although the nonlinear characteristic $n_p(x)$ will not normally be periodic it can be approximated over a given range by a Fourier sine series. Thus assuming

$$n_p(x) = \sum_{m=1}^{M} A_m \sin mcx$$

$N_p(a)$ from equation (4.11) is

$$N_p(a) = (2/a\pi) \sum_{m=1}^{M} A_m \int_0^{\pi} \sin(mca \cos\theta)\cos\theta \, d\theta$$

which on integrating gives

$$N_p(a) = (2/a) \sum_{m=1}^{M} A_m J_1(mca) \qquad (4.39)$$

where J_1 is the Bessel function of order 1.

4.5 PROPERTIES OF THE DF OF A SVNL

It is important to draw attention to some further properties of the DF of a nonlinearity. A possible general approach to selecting a quasilinear approximation for a nonlinear element is to choose a gain, K, such that some measure of the error, e, between the nonlinearity output and the linear approximation is a minimum. For the case where $x=a\cos\theta$ then

$$e = n(a\cos\theta) - Ka\cos\theta = y(\theta) - Ka\cos\theta. \qquad (4.40)$$

Choosing to minimise the mean square value σ^2 of this signal, we have

$$\sigma^2 = (1/2\pi)\int_0^{2\pi} (y(\theta) - Ka\cos\theta)^2 \, d\theta$$

which is a minimum when

$$d\sigma^2/dK = \int_0^{2\pi} 2\{y(\theta) - Ka\cos 0\}(-a\cos\theta)\, d\theta = 0$$

that is

$$K = (1/a\pi)\int_0^{2\pi} y(\theta)\cos\theta \, d\theta = N(a) . \qquad (4.41)$$

Thus the DF is also that gain which minimises the mean square error between the nonlinearity output and the output of the linear approximation.

A further interpretation for the DF is in terms of correlation functions. The crosscorrelation function between the nonlinearity input and output $R_{xy}(\tau)$ is given by

$$R_{xy}(\tau) = (1/2\pi) \int_0^{2\pi} x(\theta)y(\theta+\omega\tau)\, d\theta \qquad (4.42)$$

where τ is a time delay. Expressing $y(\theta)$ in the Fourier series

$$y(\theta) = \sum_{s=1(2)}^{\infty} a_s \cos s\theta \qquad (4.43)$$

and evaluating eq. (4.42) gives

$$R_{xy}(\tau) = (aa_1 \cos \omega\tau)/2.$$

Similarly the autocorrelation function of the input $x(\theta)$ is

$$R_{xx}(\tau) = (1/2\pi) \int_0^{2\pi} x(\theta)x(\theta + \omega\tau)\, d\theta$$

$$= (a^2 \cos \omega\tau)/2. \qquad (4.44)$$

Thus

$$\frac{R_{xy}(\tau)}{R_{xx}(\tau)} = \frac{a_1}{a} = N(a) \qquad (4.45)$$

and the DF may be considered as the ratio of the crosscorrelation between the input and output to the autocorrelation of the input. It also follows directly from the definition of the crosscorrelation function that the crosscorrelation between the error e and the nonlinearity input $x(\theta)$, namely $R_{ex}(\tau)$, is zero. A further consequence of these correlation results is that the DF provides not only the best linear gain model for the nonlinearity but also the best possible linear transfer function approximation assuming the error criterion to be that of minimum mean square.[6]

In view of the results on absolute stability presented in Chapter 3, where the sector bounds of the nonlinearity are involved, it is important to know how the value of the DF for a nonlinearity relates to the sector bounds (k_1, k_2) within which the nonlinearity is contained. Since $k_1 x < n(x) < k_2 x$ for all $0 < x < a$ it follows directly from eq. (4.5) that

$$k_1 < N(a) < k_2 \tag{4.46}$$

for all a. The DF is thus bounded by the sector gains. It can also be shown that this result is true for the magnitude of the DF of a double valued nonlinearity [7]. Also as a consequence of eq. (4.15) it follows that if $n_1(x) > n_2(x)$ for all $x > 0$, then $N_1(a) > N_2(a)$ for any a.

4.6 STABILITY DETERMINATION

If in the autonomous system of Fig. 1.1 we represent the nonlinearity by its DF then the characteristic equation of the system is

$$1 + N(a)G(s) = 0. \tag{4.47}$$

Since $N(a)$ is expected to be a good approximation for the nonlinearity for behaviour near to a limit cycle it is appropriate to examine the relationship using frequency domain methods. Here we will concentrate on Nyquist or inverse Nyquist diagrams, and root loci plots, although Bode diagrams, Nichols charts, the parameter plane and other techniques can be used. For $s = j\omega$ eq. (4.47) can be written.

$$G(j\omega) = -1/N(a) \tag{4.48}$$

so that possible a-ω solutions can be investigated by plotting $G(j\omega)$ and $-1/N(a)$, which we will denote by $C(a)$, on a Nyquist diagram or alternatively $1/G(j\omega)$ and $-N(a)$ on an inverse Nyquist diagram. If no intersections exist then the stability of the loop can be assessed by applying the Nyquist encirclement criterion to any point, C, on the $C(a)$ locus. That is, the system will be stable if the number of counterclockwise encirclements, N, of the point C is equal to the number of right half plane poles, P, of G(s). As $N(a)$ is real for a SVNL, the locus $C(a)$ always lies on the real axis of the Nyquist diagram and it is often possible to calculate analytically the parameters a–ω if the two loci intersect. This is also the case for a relay with hysteresis and no dead zone as it is easily shown from eqs. (4.34) and (4.35) with $\delta = 0$ that this particular DVNL has a $C(a)$ locus parallel to the real axis and at a distance $\pi\Delta/4h$ below it.

To illustrate the application of the DF method we present two examples.

Example 1

Consider the stability of a nonlinear system of the form of Fig. 1.1 with

$$G(s) = (1 + 0.5s)(1 + 0.1s)/\{s(s^2 + 0.5s + 2)(1 + 0.2s)\}$$

and an odd symmetric nonlinearity given by

$$n(x) = \begin{cases} x & \text{for } 0 < x \leq 1 \\ 4x - 3 & \text{for } x > 1. \end{cases}$$

The nonlinearity is easily shown to be identical to a gain of 4 and an ideal saturation characteristic with slope, $m = -3$, and saturation level, $\delta = 1$, in parallel. The $C(a)$ locus thus lies on the negative real axis from -1 to -0.25, as shown on Fig. 4.5, where the arrow denotes the direction of increasing amplitude a. The Nyquist locus, $G(j\omega)$, is also shown and it is seen to intersect the DF at C where the frequency is 1.563 rad/s. The corresponding amplitude on the $C(a)$ locus, which is defined by

$$C(a) = -\{4 - 3f_1(1/a)\}^{-1}$$

is 1.134. The root locus for G(s) is shown in Fig. 4.6. The DF method thus indicates that a limit cycle with amplitude 1.134 and frequency 1.563 rad/s. is possible. It is shown in Section 4.10 that this limit cycle is in fact unstable. Since application

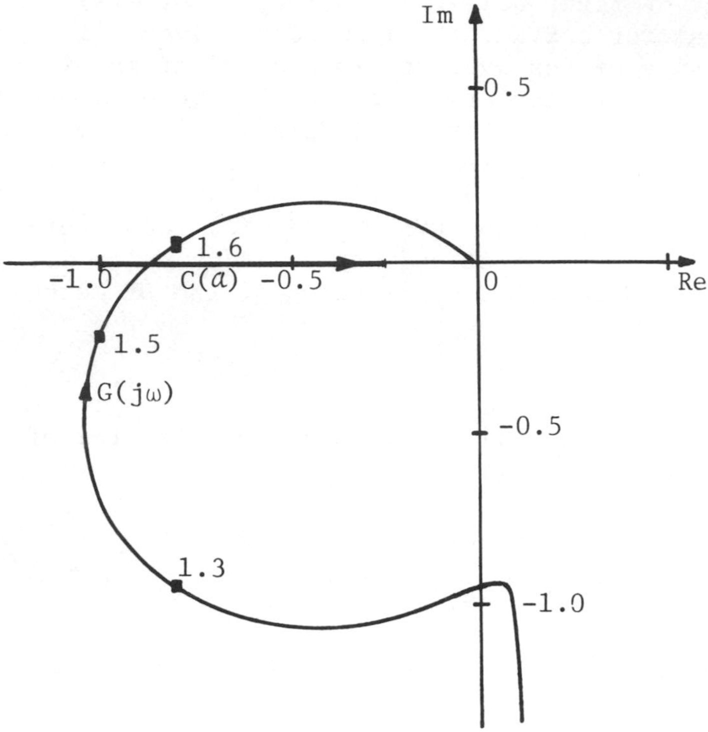

Figure 4.5 Nyquist diagram for example 1

of the Nyquist criterion to points on the $C(a)$ locus to the right of C indicates the system to be unstable for these conditions, the DF method shows that if amplitudes greater than 1.134 are reached the system output, $c(t)$, will become unbounded. Because the DF method assumes a sinusoidal input to the nonlinearity one cannot expect instability to occur exactly for initial conditions on $\dot{c}(0)$ of zero and $c(0)$ of magnitude greater than 1.134. Analog simulation of this system indicated instability for $|c(0)|>1.6$.

Example 2

Here we consider the system of Fig. 1.1 to have

$$G(s) = (1 + 0.5s)(1 + 0.1s)/\{s(s^2 + s + 1)(1 + 0.04s)\}$$

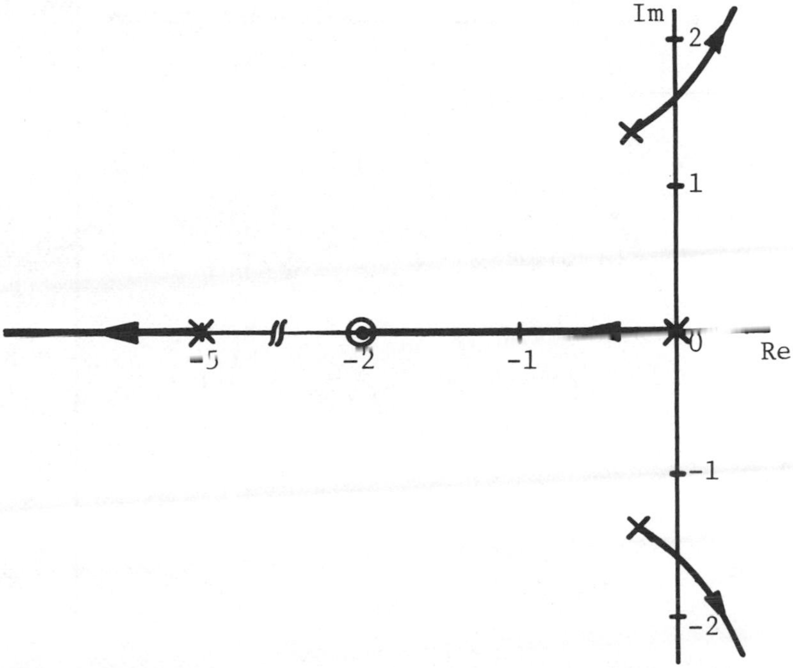

Figure 4.6 Root locus for example 1

and the relay nonlinearity of Fig. 4.3 with δ=0.7, Δ=0.3, and h=2.

Fig. 4.7 shows the relay C(a) locus and the Nyquist locus G(jω). Intersection of the two loci occurs at the point C where ω=1.26 rad/s. and a=1.61. This limit cycle is shown to be stable in Section 4.10. Any initial conditions or disturbances which cause the relay to switch will excite the limit cycle. In this case, however, for large initial conditions the limit cycle condition will be attained and the output will not become unbounded.

One aspect of the DF approach which makes it particularly attractive for engineering purposes is its value as a design aid. For example, if the system in example 2 is required to be stable with no limit cycle one can see from Fig. 4.7 how the C(a) or G(jω) loci might be modified to accomplish this specification.

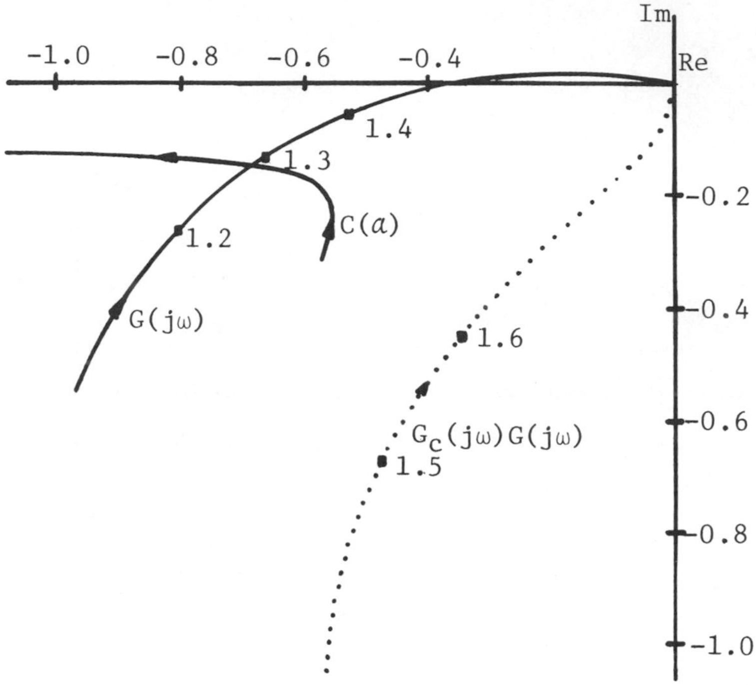

Figure 4.7 Nyquist diagram for example 2

The steady state accuracy of the system will increase as the relay dead zone $\delta+\Delta$ is decreased so presumably the value of unity taken in the example cannot be increased. It may be possible to decrease Δ but even with $\Delta=0$ and $\delta=1$ the $C(a)$ locus, which is now on the negative real axis to a maximum value of $-\pi/4$, will still intersect $G(j\omega)$. The best approach, therefore, appears to be the addition of a phase advance compensator $G_c(j\omega)$ to ensure that the new open loop frequency response $G_c(j\omega)G(j\omega)$ does not intersect $C(a)$. The resulting polar locus with $G_c(s)=(1+s)/(1+0.5s)$ is shown dotted in Fig. 4.7. From this we conclude that the system will be stable with the compensator added.

It is conventional in the design of linear systems using frequency response methods to indicate 'how stable' the system is by use of the concepts of gain and phase margin. For a nonlinear system the (-1,0), or Nyquist point, can be viewed as moving on the C(a) locus as the nonlinearity input amplitude is varied, so that a different value of phase and gain margin exists for each amplitude, a. With good computation facilities these values can be plotted against a. For most purposes, however, specification of the minimum phase and gain margins and the value of a at which they occur is a sufficient indicator of the relative stability of the system. Another approach is to plot against ω the gain ratio and phase angle curves; these are defined and evaluated as indicated in Fig. 4.8 for a selected frequency $\omega=\omega'$ and amplitude $a=a'$ to a SVNL. The region of interest is where the gain ratio is near 0 dB and the phase angle, ψ, is small, since a DF limit cycle solution corresponds to a gain ratio and phase angle of zero.

4.7 THE SINE PLUS BIAS DESCRIBING FUNCTION

For several reasons, such as the fact that constant reference inputs may exist to the feedback system

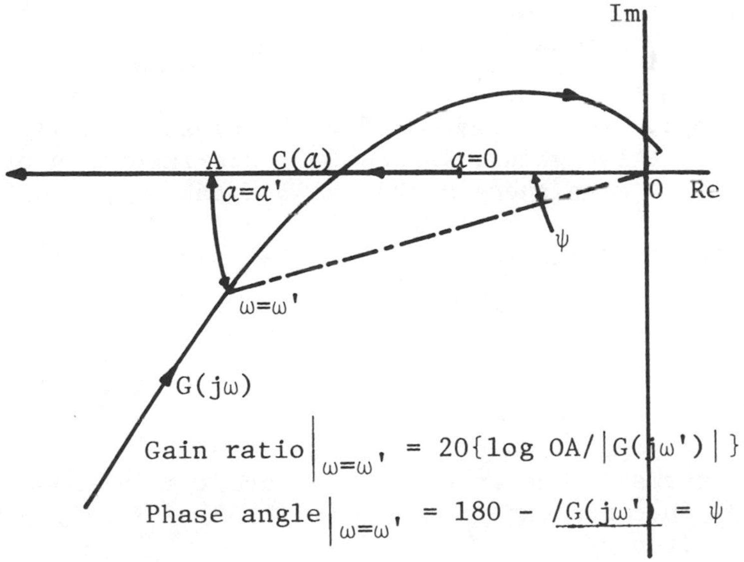

$$\text{Gain ratio}\Big|_{\omega=\omega'} = 20\{\log OA/|G(j\omega')|\}$$

$$\text{Phase angle}\Big|_{\omega=\omega'} = 180 - \angle G(j\omega') = \psi$$

Figure 4.8 Interpretation of gain ratio and phase angle

under consideration or that the nonlinearity within the loop may not be odd symmetrical, it is necessary to consider a DF for a sinusoid plus bias input to the nonlinearity, this will be denoted by SBDF. The nonlinearity input is z, where

$$z = x + \gamma \tag{4.49}$$

with $x = a \cos\theta$ the sinusoid and γ the constant bias signal. If the corresponding output waveform, $y(\theta)$, is obtained then by Fourier analysis

$$a_0 = (1/\pi)\int_0^{\pi} y(\theta)\, d\theta \tag{4.50}$$

$$a_1 = (2/\pi)\int_0^{\pi} y(\theta) \cos\theta\, d\theta \tag{4.51}$$

for the SVNL and

$$a_0 = (1/2\pi)\int_{-\pi}^{\pi} y(\theta)\, d\theta \tag{4.52}$$

$$a_1 = (1/\pi)\int_{-\pi}^{\pi} y(\theta) \cos\theta\, d\theta \tag{4.53}$$

$$b_1 = (1/\pi)\int_{-\pi}^{\pi} y(\theta) \sin\theta\, d\theta \tag{4.54}$$

for the DVNL.

The SBDF will have two component gains, one for the sinusoid, $N(a,\gamma)$, and one for the bias, $N_\gamma(a,\gamma)$. Both of these gains, as denoted, are functions of the two input magnitudes a and γ, and can be evaluated from

$$N_\gamma(a,\gamma) = a_0/\gamma \tag{4.55}$$

and

$$N(a,\gamma) = (a_1 - jb_1)/a = N_p(a,\gamma) + jN_q(a,\gamma) \tag{4.56}$$

for the general case of a DVNL.

In terms of the input signals and the in phase and quadrature nonlinearities eqs. (4.52) - (4.54) may be written as

$$a_0 = \int_{-a}^{a} n_p(x + \gamma)p(x)\, dx \tag{4.57}$$

$$a_1 = (2/a) \int_{-a}^{a} n_p(x + \gamma)xp(x)\, dx \tag{4.58}$$

and

$$b_1 = (2/a\pi) \int_{-a}^{a} n_q(x + \gamma)\, dx. \tag{4.59}$$

Table I.2 in Appendix I contains a short table of SBDFs for odd symmetric nonlinearities.

For the SVNL it is easy to show that the optimum linear model of the nonlinearity for a minimum mean square error is $aN(a,\gamma) + \gamma N_\gamma(a,\gamma)$ and also that

$N(a,\gamma)=R_{xy}(\tau)/R_{xx}(\tau)$ and $N_\gamma(a,\gamma)=R_{\gamma y}(\tau)/R_{\gamma\gamma}(\tau)$ in

agreement with extensions of the properties given in Section 4.5 for the DF.

A function of interest for stability investigations is the dc incremental gain $g(a,\gamma)$ given by

$$g(a,\gamma) = da_0/d\gamma \tag{4.60}$$

which in the limit as $\gamma\to 0$ gives the gain to any small signal unrelated to the sinusoid, x. This gain, known as the incremental describing function, IDF, and denoted by $N_{i\gamma}(a)$ is given, for a SVNL, by

$$N_{i\gamma}(a) = g(a,0) = \lim_{\gamma\to 0} \frac{d}{d\gamma} \int_{-a}^{a} n(x + \gamma)p(x)\, dx$$

$$= \int_{-a}^{a} n'(x)p(x)\, dx \tag{4.61}$$

where $n'(x)=d\{n(x)\}/dx$. Integrating eq. (4.5), with the integral expressed over the range $-a$ to a, by parts gives

$$N(a) = (2/a^2) \int_{-a}^{a} (a^2 - x^2)n'(x)p(x)dx \tag{4.62}$$

so that

$$a^2N(a) = (2/\pi) \int_{-a}^{a} n'(x)(a^2 - x^2)^{1/2}\, dx. \tag{4.63}$$

Differentiating both sides with respect to a gives

$$2aN(a) + a^2\,(dN(a)/da) = 2\int_{-a}^{a} an'(x)p(x)\, dx.$$

Substituting for the right hand side integral in eq. (4.61) gives

$$N_{i\gamma}(a) = N(a) + (a/2)N'(a) \tag{4.64}$$

where $N'(a)=dN(a)/da$, and shows that $N_{i\gamma}(a)$ can be found from $N(a)$.

4.8 THE GENERAL IDF

The effect of adding a small perturbation to a limit cycle has to be investigated in order to assess the stability of the limit cycle. The results obtained for the SBDF were used in the previous section to derive an expression for the IDF for the particular case where the incremental input γ is unrelated to the sinusoid, x. In general, however, it is not sufficient to consider an unrelated perturbation. To evaluate a general IDF we thus have to consider all possible perturbations. Let $\delta(t)$ be a small perturbation in the presence of the sinusoid, x, then the nonlinearity output may be approximated by the first two terms of a Taylor series, that is

$$n\{x(t) + \delta(t)\} \simeq n\{x(t)\} + \delta(t)n'\{x(t)\} \tag{4.65}$$

where the additional output due to the increment $\delta(t)$ is $\delta(t)n'\{x(t)\}$. For an input $x=a \cos \theta_a$ the output of the nonlinearity $n'(x)$ will be periodic and can be represented for any nonlinearity by the Fourier series

$$n'\{x(t)\} = \sum_{s=0}^{\infty} c'_s \cos (s\theta_a + \phi'_s) , \tag{4.66}$$

with $\phi'_0=0$. Thus if $\delta(t)=b \cos (\theta_b + \phi)$,

$$\delta(t)n'\{x(t)\} = b \sum_{s=0}^{\infty} c'_s \cos (s\theta_a + \phi'_s) \cos (\theta_b + \phi)$$

$$= b \sum_{s=0}^{\infty} (c'_s/2) \{\cos (s\theta_a + \theta_b + \phi'_s + \phi) + \cos (s\theta_a - \theta_b + \phi'_s - \phi)\} . \tag{4.67}$$

If $x(t)$ and $\delta(t)$ are unrelated then the only output at frequency ω_b, where $\theta_b=\omega_b t$, is $c_0'b \cos(\theta_b+\phi)$ obtained with $s=0$. On the other hand, if $\omega_a/\omega_b=m/n$, where m and n are integers, then an additional output at frequency ω_b, for s equal to $2n/m$ occurs in eq. (4.67) and is given by $(c_s'b/2) \cos(\theta_b+\phi_s'-\phi)$. The total output at frequency ω_b is thus;

$$c_0'b \cos(\theta_b + \phi) + \{(c_s'b/2) \cos(\theta_b+\phi_s'-\phi)\}\big|_{s=2n/m}.$$

The gain at this frequency, $N_{ib}(a)$, is then given by

$$N_{ib}(a) = \{c_0' + (c_s'/2)e^{j(\phi_s' - 2\phi)}\}_{s=2n/m}. \quad (4.68)$$

The case of primary interest is where n=m, for which s=2, and

$$N_{ib}(a) = c_0' + (c_2'/2)e^{j(\phi_2' - 2\phi)}. \quad (4.69)$$

For a SVNL, $\phi_2'=0$,

$$c_0' = \int_{-a}^{a} n'(x)p(x)dx = N_{i\gamma}(a), \quad (4.70)$$

and in addition it can be shown that (see Section 5.2)

$$c_2' = 2\int_{-a}^{a} n'(x)\{(2x^2/a^2) - 1\}p(x)\, dx, \quad (4.71)$$

which using eq. (4.62) gives

$$c_2' = 2\{N_{i\gamma}(a) - N(a)\}. \quad (4.72)$$

Using these results in eq. (4.69) and the expression for $N_{i\gamma}(a)$ from eq. (4.64) gives for a SVNL;

$$N_{ib}(a) = N(a) + (a/2)N'(a)(1 + e^{-j2\phi}). \quad (4.73)$$

This is the equation of a circle with centre at the point $\{N_{i\gamma}(a), 0\}$ and radius $(a/2)N'(a)$.

4.9 ASYMMETRICAL OSCILLATIONS

Fig. 4.9 shows the block diagram of a system with a constant reference input, R, and a constant

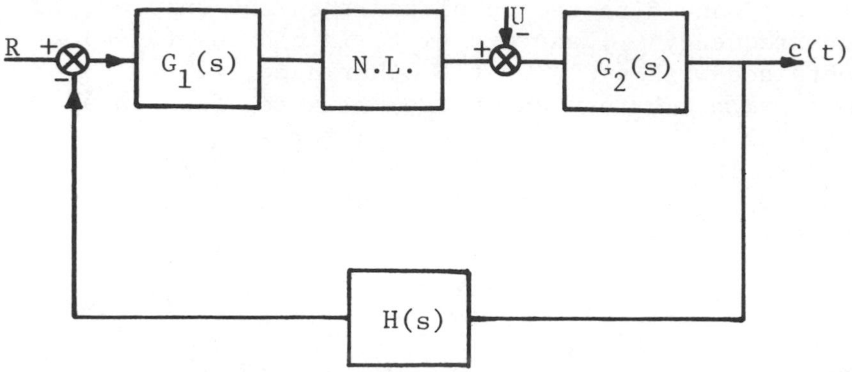

Figure 4.9 System block diagram

disturbance input, U. For a limit cycle to exist in the loop two conditions are to be satisfied, one as previously for the fundamental frequency, since higher harmonics will again be neglected, and the other for the bias signals. Because superposition applies for the linear elements, the resulting equations are

$$1 + G(j\omega)N(a,\gamma) = 0 \tag{4.74}$$

and

$$\gamma\{1 + N_\gamma(a,\gamma)G(0)\} = RG_1(0) + UG(0), \tag{4.75}$$

where $G(s)=G_1(s)G_2(s)H(s)$. Eq. (4.75) may be simplified if an integrator or differentiator exists in the loop, since in the steady state the bias input to the former and output of the latter must be zero. The graphical approach analogous to that given in Section 4.6 for $\gamma=0$ is to assume values for a in eq. (4.75) and solve for the corresponding values of γ, say $\gamma(a)$. $-1/N(a,\gamma(a))$ then becomes the $C(a)$ locus plotted on the Nyquist diagram to determine possible solutions to eq. (4.74). If for a given value of a there are two solutions for $\gamma(a)$ from eq. (4.75) then two $C(a)$ loci must be plotted. The parameters of a limit cycle can then be found from any intersection of $C(a)$ and $G(j\omega)$, or if no intersection exists the stability of the system can be considered using the Nyquist criterion with respect to any point on the

C(a) locus. The stability of a predicted limit cycle is considered in the next section.

It is worth noting here, although the problem is considered further in the next chapter, that the DF method usually produces less accurate results in asymmetrical systems. This is simply explainable by the filtering considerations put forward in the introduction to the chapter. Because the oscillation waveform at the nonlinearity output will now contain even, and in particular some second, harmonics, the distortion in the signal fed back to the nonlinearity will in general be greater than for the odd symmetrical case.

As a specific example of the above theory we consider the loop of Fig. 4.9 with the nonlinearity an ideal relay, $G_1(s) = 1/(s + 1)$, $G_2(s) = K/s(s + 1)$ and $H(s) = 1$.

From eqs. (4.57) and (4.58) we obtain for the ideal relay, for $|\gamma|<a$,

$$a_0 = \int_{-a}^{-\gamma} -hp(x)\,dx + \int_{-\gamma}^{a} hp(x)\,dx$$

and

$$a_1 = (2/a)\left\{\int_{-a}^{-\gamma} -hxp(x)\,dx + \int_{-\gamma}^{a} hxp(x)\,dx\right\}.$$

Evaluating the integrals gives;

$$a_0 = (2h/\pi)\sin^{-1}(\gamma/a) \qquad (4.76)$$

and

$$(a_1/a) = N(a,\gamma) = (4h/a^2\pi)(a^2 - \gamma^2)^{1/2}. \qquad (4.77)$$

The harmonic and bias loop equations, corresponding to eqs. (4.74) and (4.75), are

$$-1/N(a,\gamma) = K/j\omega(1 + j\omega)^2 \qquad (4.78)$$

and

$$U = a_0 = \gamma N_\gamma(a,\gamma) \qquad (4.79)$$

the latter being simplified because of the integrator in $G_2(s)$. In addition if the average value, C, of c(t) is required this can be obtained from

$$C = R - \gamma. \tag{4.80}$$

Since $N(a,\gamma)$ is real and positive the $C(a)$ locus is along the negative real axis and any limit cycle frequency is given by $\underline{/G(j\omega)}=180°$, that is

$$90 + 2\tan^{-1}\omega = 180°$$

which gives $\omega=1$ rad/s. and $|G(j\omega)|_{\omega=1}=K/2$
Using the expressions for a_0 and $N(a,\gamma)$ from eqs. (4.76) and (4.77) thus gives for a limit cycle

$$U = (2h/\pi)\sin^{-1}(\gamma/a)$$

and

$$K/2 = (4h/a^2\pi)(a^2 - \gamma^2)^{1/2}$$

which on solving for γ and a yields

$$\gamma = (hK/\pi)\sin(U\pi/h) \tag{4.81}$$

and

$$a = (2hK/\pi)\cos(U\pi/2h). \tag{4.82}$$

From eq. (4.81) the maximum value of γ is hK/π when $U/h=1/2$. This means that a DF solution for the limit cycle only exists for $|U|<h/2$, a solution error caused by the inaccuracy of the DF as the distortion increases. The fundamental component, a, of the limit cycle at the nonlinearity input decreases as U increases and has a value approaching 0.707 times its value for U=0 when the oscillation ceases. The frequency of oscillation remains constant at 1 rad/s. as U is varied according to the above DF solution. The exact analysis method presented in Chapter 6 shows that the actual frequency diverges farther from 1 rad/s. as U is increased.

This is to be expected as the percentage distortion in c(t) will increase, due primarily to the second harmonic, as the relay operation becomes more asymmetric.

4.10 LIMIT CYCLE STABILITY

When eq. (4.48) yields a solution for a limit cycle with $a=a_o$ and $\omega=\omega_o$ we wish to know if the limit cycle is stable. To investigate the stability of a periodic solution we need to consider what happens when the solution is subjected to a small perturbation. Known results suggest two forms of perturbation should be considered, one where the perturbation is at the solution frequency, corresponding to a small change in the limit cycle parameters, and the other where the perturbation is at any other frequency. This is to account for the fact that if the predicted limit cycle existed it would change the gain through the nonlinearity to any other signal. It is usually sufficient to consider this perturbation to be unrelated to the oscillation.

Considering first a limit cycle perturbation we may write eq. (4.48) in the form

$$X(a,\omega) + jY(a,\omega) = 0 \qquad (4.83)$$

and assume it has a limit cycle solution $a=a_o \exp(j\omega_o t)$. Neglecting any effects of the rate of change of frequency and damping [8] and taking a small perturbation in a_o to $a_o+\Delta a$ and in ω_o to $\omega_o+\Delta\omega+j\Delta\sigma$, the perturbed value of a becomes $(a_o+\Delta a)\exp\{j(\omega_o+\Delta\omega+j\Delta\sigma)t\}$. If $\Delta\sigma$ is positive the real part of the exponential is negative and a is positively damped (i.e. decaying). Substituting in eq. (4.83) gives

$$X(a_o+\Delta a,\ \omega_o+\Delta\omega+j\Delta\sigma) + jY(a_o+\Delta a, \omega_o+\Delta\omega+j\Delta\sigma) = 0$$

which on expanding in a Taylor series to first order terms about the equilibrium state (a_o,ω_o) and removing the limit cycle solution equation $X(a_o,\omega_o) + jY(a_o,\omega_o) = 0$ gives

$$\left.\frac{\partial X}{\partial a}\right|_s \Delta a + \left.\frac{\partial X}{\partial \omega}\right|_s \Delta\omega + j\left.\frac{\partial X}{\partial \omega}\right|_s \Delta\sigma + j\left.\frac{\partial Y}{\partial a}\right|_s \Delta a + j\left.\frac{\partial Y}{\partial \omega}\right|_s \Delta\omega -$$

$$- \left.\frac{\partial Y}{\partial \omega}\right|_s \Delta\sigma = 0 ,$$

where $|_s$ denotes that the partial derivative is evaluated at the solution (a_o,ω_o). Separating the equation into its real and imaginary parts gives

$$\left.\frac{\partial X}{\partial a}\right|_s \Delta a + \left.\frac{\partial X}{\partial \omega}\right|_s \Delta\omega - \left.\frac{\partial Y}{\partial \omega}\right|_s \Delta\sigma = 0 ,$$

and

$$\left.\frac{\partial Y}{\partial a}\right|_s \Delta a + \left.\frac{\partial Y}{\partial \omega}\right|_s \Delta\omega + \left.\frac{\partial X}{\partial \omega}\right|_s \Delta\sigma = 0 .$$

Eliminating $\Delta\omega$ from these equations one has

$$\left.\left(\frac{\partial X}{\partial a}\frac{\partial Y}{\partial \omega} - \frac{\partial X}{\partial \omega}\frac{\partial Y}{\partial a}\right)\right|_s \Delta a = \left| \left.\left(\frac{\partial X}{\partial \omega}\right)\right|_s^2 + \left.\left(\frac{\partial Y}{\partial \omega}\right)\right|_s^2 \right| \Delta\sigma .$$

For the limit cycle to be stable a positive Δa must lead to a positive $\Delta\sigma$ and vice versa, so that a necessary condition for the limit cycle to be stable is

$$\left.\left(\frac{\partial X}{\partial a}\frac{\partial Y}{\partial \omega} - \frac{\partial X}{\partial \omega}\frac{\partial Y}{\partial a}\right)\right|_s > 0 . \tag{4.84}$$

Although this is a suitable analytical result it is often easier to apply if it is extended to allow a graphical interpretation. Taking

$$G(j\omega) = U(\omega) + jV(\omega) ,$$

and

$$C(a) = -1/N(a) = P(a) + jQ(a) ,$$

then eq. (4.83) becomes

$$G(j\omega) + \{1/N(a)\} = U(\omega) + jV(\omega) - P(a) - jQ(a) ,$$

so that

$$X(a,\omega) = U(\omega) - P(a) \quad \text{and} \quad Y(a,\omega) = V(\omega) - Q(a) .$$

Substituting in eq. (4.84), the necessary condition for a stable limit cycle becomes

$$\left.\left(\frac{\partial U}{\partial \omega}\frac{\partial Q}{\partial a} - \frac{\partial V}{\partial \omega}\frac{\partial P}{\partial a}\right)\right|_s > 0 , \qquad (4.85)$$

which, if N(a) is real, reduces further to

$$\left.\left(\frac{\partial V}{\partial \omega}\frac{\partial P}{\partial a}\right)\right|_s < 0 . \qquad (4.86)$$

Examination of the left hand side of eq. (4.86) shows that it is equal to the vector product $\partial G/\partial\omega \times \partial C/\partial a$, which may be denoted by g x c. The necessary condition for a stable limit cycle thus requires the vectors (g,c,v), where v=gxc, to form a right hand triple at the solution point. This result is known as the Loeb criterion [9].

A similar graphical criterion may also be arrived at by logical argument under the same conditions that apply to the analytical solution. For a stable limit cycle, if a small increase in amplitude along the C(a) locus from the solution is considered, then the movement should be into the region of a stable system configuration and for a small decrease in amplitude along the C(a) locus, the movement should be into the region of an unstable system configuration. In both cases the perturbation moves the system into a state which will counteract the perturbation so as to restore the system to the limit cycle solution. The locations of the loci $G(\Delta\sigma+j\omega)$ and $G(-\Delta\sigma+j\omega)$ with respect to the $G(j\omega)$ locus, where $\Delta\sigma$ is a small positive quantity, are easily found from the rules of the s to G(s) plane mapping.

Consider, for example, the situation portrayed in Fig. 4.10 where eq. (4.48) is shown to have three solutions at the points A, B and C respectively. If the transfer function G(s) has no right half plane poles then the σ positive and σ negative regions will be as shown in the figure. For small positive increments in a along $C(a)$ from the solution points, movement is into regions where $\Delta\sigma$ is negative, positive and negative respectively and vice versa for decrements in a. Thus since the Loeb criterion is necessary for stability we may conclude that the limit cycle at B is unstable and those at A and C may be stable.

Possible limit cycle frequencies may also be obtained from where the root loci of G(s) cross the imaginary axis when the system contains a SVNL, since its DF is real. The crossover frequency will be a valid DF solution for the system concerned if a value a_o for a can be found such that the loop gain $KN(a_o)$, assuming the linear transfer function is KG(s), can be made equal to the root locus gain at the crossover point. Further, it can be shown that the Loeb criterion may be applied to the root locus plot to assess the stability of these limit cycle solutions. The necessary condition for a limit cycle to be stable is that the crossover of the imaginary axis be from left (right) to right (left) if $N'(a_o)$ is negative (positive). The limit cycle indicated by the root locus plot of Fig. 4.6 for example 2 in Section 4.6 is thus unstable, since the DF for the nonlinearity used has $N'(a_o)$ positive.

The advantages of the Loeb criterion and the perturbation of the solution approach discussed above are the ease with which they can be applied to DF solutions obtained graphically. For transfer functions G(s) which give multiple DF limit cycle solutions the perturbation solution approach may be invalid, since it is possible, for example, to choose a $G(j\omega)$ such that a perturbation in either direction from the solution point may be into an unstable region yet the limit cycle predicted may be stable.

The differential equation describing the system of Fig. 1.1 can be written

$$q(D)x(t) + p(D)n\{x(t)\} = 0 \qquad (4.87)$$

where D denotes the differential operator and $G(s) = p(s)/q(s)$. Assuming a limit cycle solution with

$x(t)=x^*(t)$, then for a small perturbation in $x^*(t)$ of $\Delta x(t)$ we have

$$q(D)\{x^*(t) + \Delta x(t)\} + p(D)n\{x^*(t) + \Delta x(t)\} = 0$$

Expanding the nonlinear function in a Taylor series and using the limit cycle solution equation, that is eq. (4.87) with $x(t)=x^*(t)$, gives the variational equation

$$q(D)\Delta x(t) + p(D)n'\{x^*(t)\}\Delta x(t) = 0 \qquad (4.88)$$

This equation is a linear differential equation with periodically time varying coefficients due to the term $n'\{x^*(t)\}$. Two difficulties are now encountered. Firstly even if the exact solution for $x^*(t)$ is available no necessary and sufficient conditions for the stability of eq. (4.88) are known; although when the equation is of second order the stability is the same as that of the time averaged variational equation [10]. It is easily shown that

$$\overline{n'\{x^*(t)\}} = N_{i\gamma} \qquad (4.89)$$

where — denotes a time average and $N_{i\gamma}$ is the d.c. incremental gain of $n(x)$ with $x^*(t)$ as input, so that the time averaged variational equation is

$$\{q(D) + p(D)N_{i\gamma}\}\Delta x(t) = 0 \qquad (4.90)$$

The second point is that in using the DF method we find an approximation $a_o \sin \omega_o t$ to $x^*(t)$ and hence can only approximate $N_{i\gamma}$ by $N_{i\gamma}(a_o)$. For this case eq. (4.90) becomes

$$\{q(D) + p(D)N_{i\gamma}(a_o)\}\Delta x(t) = 0 \qquad (4.91)$$

and its stability is given by the roots of the characteristic equation

$$1 + N_{i\gamma}(a_o)G(s) = 0 \qquad (4.92)$$

If the perturbation is considered to be synchronous the corresponding characteristic equation is

$$1 + N_{ib}(a_o)G(\sigma+j\omega_o) = 0 \qquad (4.93)$$

Since $N_{ib}(a_o)$ is equal to $N(a)$ for $\sigma=\pi/2$ two solutions of this equation are the limit cycle solution with $\sigma=0$. The number of additional solutions to the equation is not easily determined, although it can be shown [11] that at least one other solution will exist for σ negative, if the Loeb criterion indicates a stable limit cycle, and for σ positive, if the Loeb criterion predicts an unstable limit cycle. Evaluation of the roots for several examples has produced no situation of roots with σ positive where the Loeb necessary condition for a stable limit cycle is satisfied. In summary eq. (4.93) appears to give results in agreement with the Loeb criterion.

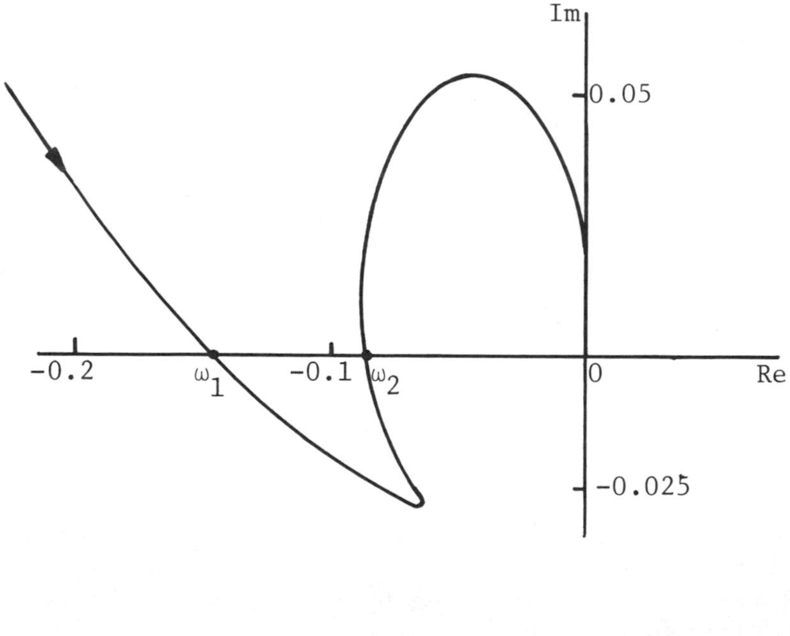

Figure 4.10 $G(j\omega)$ locus for
$G(s) = (s+1)^2/s^3(s^2+1.05s+12.25)$

When the root locus of G(s) has more than one branch in the right hand side of the s plane it is possible for eq. (4.92) to have a root with a positive real part, so that the predicted limit cycle is unstable, when the Loeb necessary condition for a stable limit cycle is satisfied. Because the DF solution is approximate no criterion for the stability of a DF predicted limit cycle solution will give correct results in all cases. Indications are that when the roots of eq. (4.92) give incorrect stability information the error is due to inaccuracies in the DF solution. This approach can also be used for asymmetrical systems provided the IDF in eq. (4.92), which will be $N_{i\gamma}(a_o, \gamma_o)$, is evaluated at the operating point (a_o, γ_o).

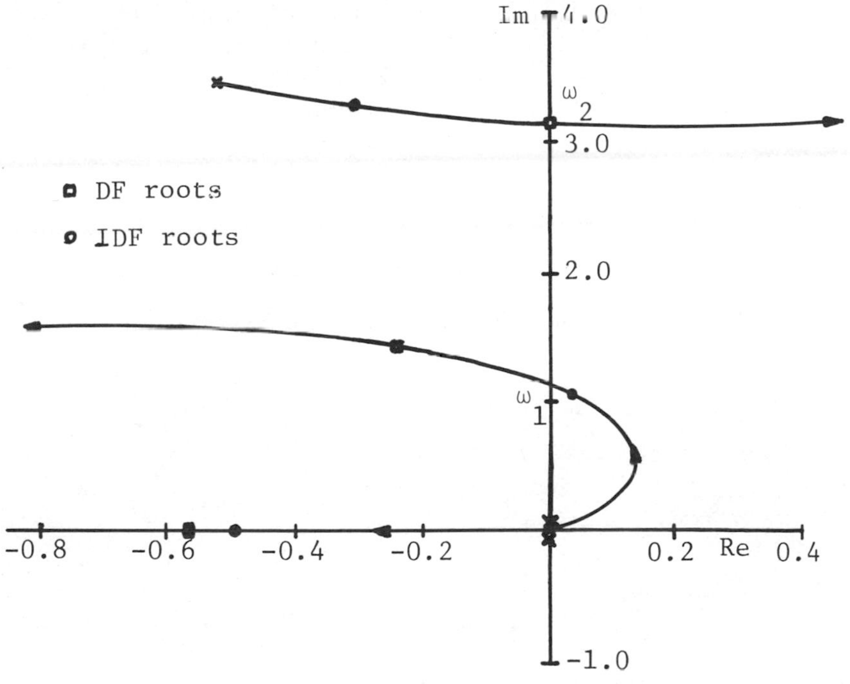

Figure 4.11 Root locus for G(s) of Fig. 4.10

An example illustrating the above points is given in reference [11]. The nonlinearity is an ideal relay and $G(s)=(s+1)^2/\{s^3(s^2+1.05s+12.25)\}$. Fig. 4.10 shows the Nyquist locus $G(j\omega)$ and Fig. 4.11 the positive frequency portion of the root locus of G(s). The poles of eq. (4.88), which are easily located since $N_{i\gamma}(a)=2/a\pi=N(a)/2$ for the ideal relay, are as shown in Fig. 4.11. Since a pole exists in the right hand side of the s plane the limit cycle at $\omega=\omega_2$ is unstable. This result was confirmed by simulation where a combined mode oscillation was found.

4.11 SUMMARY

The basic procedures for investigating the stability of a nonlinear feedback loop using the DF method have been presented in this chapter. Techniques for evaluating DFs, especially SBDFs for asymmetrical nonlinearities, have been covered briefly and for additional information the reader may refer to references [6] and [12]. The IDF approach for assessing the stability of a limit cycle predicted by the DF method has been included and its relationship to the Loeb criterion discussed. Several examples have been given to illustrate the application of the techniques presented.

REFERENCES

1. Tustin, A.: "The effects of backlash and of speed dependent friction on the stability of closed cycle control systems", J.I.E.E., 1947, pt. II, 94, pp. 143-151.

2. Dutilh, J.: "Theorie des servomechanism a relais". Onde Elec., 1950, pp. 438-445.

3. Oppelt, W.: "Locus curve method for regulators with friction", Z. Deut. Ingr., Berlin, 1940, 90.

4. Goldfarb, L.C.: "On some nonlinear phenomena in regulatory system", Automatika i Telemekhanika, 1947, pp. 349-383, Translation: R. Oldenburger (ed.), "Frequency Response", The Macmillan Co., New York, 1956.

5. Kochenburger, R.J.: "A frequency response method for analysing and synthesizing contactor servomechanisms", Trans A.I.E.E., 1950, 69, pp. 270-283.

6. Atherton, D.P.: "Nonlinear control engineering", Van Nostrand Reinhold, London, 1975, Chapters 3-5.

7. Cook, P.A.: "Describing function for a sector nonlinearity", Proc. I.E.E., 1973, 120, pp. 143-144.

8. Gelb, A. and Vander Velde, W.E.: "On limit cycling control systems", I.E.E.E. Trans., 1963, AC-8, pp. 142-157.

9. Loeb, J.M.: "Recent advances in nonlinear servo theory", in R. Oldenburger (ed.), Frequency response, The Macmillan Co., New York, 1956, pp. 260-268.

10. Willems, J.L.: "The stability of oscillations in nonlinear networks", IEEE Trans. Circuit Theory (Corresp.), vol. CT-15, 1968, pp. 284-286.

11. Choudhury, S.K. and Atherton, D.P.: "Limit cycles in high order nonlinear systems", Proc. I.E.E., 1974, 121, pp. 717-724.

12. Gelb, A. and Vander Velde, W.E.: "Multiple input describing functions and nonlinear system design", McGraw-Hill, New York, 1968, Chapters 2 and 3.

PROBLEMS

1. Prove that the $C(a)$ locus for a relay with $\delta=0$ is parallel to the real axis.

2. Show that the maximum value of $N(a)$ for a relay with $\Delta=0$ is $2h/\pi\delta$.

3. Find the maximum value of $M(a)$ for the general relay of Fig. 4.3.

4. Derive the result for $N(a)$ of the quantised nonlinearity by assuming it to be a parallel combination of relays with $\Delta=0$.

5. Show that the DF of n nonlinearities in parallel is the sum of their individual DFs.

6. Determine $N(a)$ for a linear segmented nonlinearity, $n(x)$, with $n(0)=0$, slope m_1 for x from 0 to δ_1, m_2 from δ_1 to δ_2 and m_3 beyond δ_2.

7. Find the DF of a friction controlled backlash characteristic, a symmetrical odd DVNL, with

$$n_1(x) = \begin{cases} -(a - b) & \text{for } -a \leq x < -(a - 2b) \\ x - b & \text{for } -(a - 2b) < x \leq a \end{cases} .$$

8. Find the DF of a symmetrical odd DVNL defined by

$$n_1(x) = \begin{cases} -1 & x \leq 0 \\ x-1 & 0 < x < 1 \\ 1 & x \geq 1 \end{cases} .$$

9. A position control system with negligible viscous friction and coulomb friction equivalent to an error of d radians is stabilised by velocity feedback. The velocity feedback signal is inadvertently connected in the wrong sense so that the equation of motion is governed by

$$(1/\omega^2)\ddot{c} - (2\zeta/\omega_o)\dot{c} + c + d\,\mathrm{sgn}(\dot{c}) = 0 .$$

Determine the amplitude and frequency of any possible limit cycles for $0 < \zeta < 1$, using
(a) the phase plane
(b) the describing function.
Is the limit cycle stable? Compare the results for $\zeta = 0.25$.

10. Determine the maximum allowable value of T_d for stability of a system with an ideal saturation nonlinearity and

$$G(s) = 10e^{-sT_d}/s(s+2).$$

If $T_d = 0.5$ s. determine the frequency, amplitude at the input to the nonlinearity and stability of any possible oscillation. Check your results using the Routh criterion and the Padé approximation

$$e^{-x} = \frac{1 - (3/4)x + (1/4)x^2 - (1/24)x^3}{1 + (1/4)x}.$$

11. A closed loop system contains a relay with $\Delta=0$ and a linear transfer function

$$G(s) = K\omega_o^2/s(s^2 + 2\zeta s\omega_o + \omega_o^2).$$

Show that the system will be stable if

$$0 < K < (\pi\zeta\omega_o\delta)/h.$$

If $K=3\pi\zeta\omega_o\delta/h$ find the amplitude and frequency of the limit cycle.

12. Using the describing function discuss the stability of the differential equation.

$$\dddot{x} + 2\ddot{x} + \dot{x} - (\dot{x}^3/2) + x = 0.$$

13. Using the describing function discuss the stability of the differential difference equation

$$\ddot{x}(t) - \mu\{1 - \dot{x}^2(t)\}\dot{x}(t) + x(t - 2) = 0.$$

14. A nonlinear system consists of an ideal relay and a linear transfer function

$$Ke^{-sT_d}/s(s + 2).$$

Determine all the possible frequencies of oscillation and show using the IDF that only the one at the lowest frequency is stable.

15. Show that for a nonlinearity n(x) with n'(x) monotonic decreasing

$$N_{i\gamma}(a) < n(a)/a < N(a)$$

and

$$N_{i\gamma}(a) > N(a)/2 .$$

16. Show that $-1/N_{ib}(a)$ is a circle for a given a with the equation

$$[x + \{(K_1 + K_2)/2K_1K_2\}]^2 + y^2 = \{(K_1 - K_2)/2K_1K_2\}^2$$

where $K_1 = N(a)$ and $K_2 = N(a) + aN'(a)$.

17. An ideal relay switches power from the mains to a 10 kW oven heater. The transfer function relating oven temperature θ°C above ambient, and input power W watts to the heater, is

$$\frac{\theta(s)}{W(s)} = \frac{0.05}{(1 + sT_1)(1 + sT_2)(1 + sT_3)} .$$

If $T_1 = 2T_2 = 4T_3 = 8$ secs. determine the peak amplitude and the frequency of the oscillations in temperature about a controlled level of 200°C above ambient.

18. Draw a block diagram in the form of Fig. 1.1 to represent the Raleigh equation

$$\ddot{x} + \mu\{(\dot{x}^3/3) - \dot{x}\} + x = 0 .$$

Obtain the DF solution for the amplitude and frequency of the limit cycle.

19. Obtain the DF solution for the limit cycle of Example 3, Section 2.7 for K=1, T_d=1 and λ=0.8. Compare the frequency with that obtained by the phase plane method.

20. Use the DF method to evaluate the limit cycle obtained in Example 1, Section 2.7.

21. Evaluate the SBDF and asynchronous IDF, $N_{i\gamma}$, for a relay with dead zone (i.e. Δ=0).

22. Use the IDF, $N_{i\gamma}$, to assess the stability of oscillations in a system with a SVNL n(x) and $G(s)=K/s(s+1)(s+2)$. Interpret your results on a root locus diagram if n(x) is (a) a hard spring and (b) a soft spring nonlinearity. Determine the condition for a stable oscillation if n(x) is a relay with dead zone (i.e. Δ=0).

CHAPTER 5

Additional Considerations of the Describing Function Method

5.1 INTRODUCTION

The aim of this chapter is to examine in more detail various aspects of stability investigations using the DF method. Because the technique is approximate it is important to have mechanisms for determining whether the results it produces can be relied upon. Obviously, the accuracy of the approach depends on the validity of the initial assumption that the nonlinearity input is a sinusoid. In the next section we therefore briefly discuss the evaluation of the harmonics at the output of a nonlinearity with a sinusoidal input and also the describing function for two sinusoidal inputs to a nonlinearity, to illustrate the effect of neglecting the harmonics on the gain to the fundamental. Later in Sections 5.8 and 5.9 these results will be used again when we consider the more accurate determination of limit cycle solutions assuming the nonlinearity input to be a sinusoid plus a harmonic and also when combined oscillations are discussed.

Section 5.3 discusses techniques for estimating the reliability of the DF method using both empirical and exact procedures. Aizerman [1] and Kalman [2] introduced interesting conjectures regarding the stability of the system of Fig. 1.1 assuming $n(x) \in (k_1, k_2)$ and $n'(x) \in (m_1, m_2)$. We examine these conjectures, which are not true in general, and their relationship to the DF in Section 5.4. Several counterexamples to these conjectures, which also correspond to incorrect DF predictions have been presented in the literature and some of these are discussed in Section 5.5.

Section 5.6 is devoted to studies of the stability of a few unusual examples which allow us to make use of DF concepts. In many practical problems the nonlinear effects in a system cannot be separated into a single static nonlinearity and linear dynamics. The DF method for frequency dependent nonlinearities and systems with multiple nonlinear elements is therefore considered in Section 5.7.

5.2 HARMONIC RESPONSE

The harmonic terms at the output of a nonlinearity with an input consisting of a bias signal, γ, and sine wave can be evaluated by Fourier analysis of the output waveform, $y(\theta)$. If the sinusoid is taken as $a \cos \theta$ it can also be shown [3] that the coefficients of the Fourier series of the output

$$y(\theta) = \sum_{s=0}^{\infty} a_s \cos s\theta + b_s \sin s\theta \qquad (5.1)$$

can be found from

$$a_s = \varepsilon_s \int_{-a}^{a} n_p(x + \gamma) T_s(x/a) p(x)\, dx \qquad (5.2)$$

and

$$b_s = 2 \int_{-a}^{a} n_q(x + \gamma) D_s(x/a) p(x)\, dx \qquad (5.3)$$

where $p(x)$, $n_p(x)$ and $n_q(x)$ are as defined in eqs. (4.6), (4.13) and (4.14) respectively.

ε_s is the Neumann factor, that is,

$$\varepsilon_s = \begin{cases} 1 & \text{for } s = 0 \\ 2 & \text{for } s \geq 1 , \end{cases} \qquad (5.4)$$

$D_x(s)$ is given by

$$D_s(x) = (1 - x^2)^{1/2}\, U_{s-1}(x) \qquad (5.5)$$

and $T_s(x)$ and $U_s(x)$ are Chebyshev polynomials, which can be defined by the recursion relationship

$$T_{s+1}(x) = 2xT_s(x) - T_{s-1}(x) , \qquad (5.6)$$

with the first two polynomials $T_0=1$, $T_1=x$ and $U_0=1$, $U_1=2x$.

The coefficients a_s and b_s are functions of both a and γ which we will note by writing

$$a_s = \varepsilon_s \, n_p(a,\gamma)_s \qquad (5.7)$$

and

$$b_s = n_q(a,\gamma)_s \quad . \qquad (5.8)$$

When plotted for a fixed a as γ varies $n_p(a,\gamma)_s$ and $n_q(a,\gamma)_s$ may be viewed as modified nonlinearities. This concept of a modified nonlinearity is useful for determining the output of a nonlinearity when more than one sinusoid is applied to the input. Here we restrict considerations to a SVNL and consider first the case of two sinusoidal inputs, $x_1=a \cos \theta_a$ and $x_2=b \cos (\theta_b+\theta)$, for which case it can be shown [3] that the nonlinearity output, $y(\theta)$, may be written as the double Fourier series

$$y(\theta) = \sum_{s=0}^{\infty} \sum_{k=0}^{\infty} \varepsilon_s \varepsilon_k \alpha_{sk} \cos s\theta_a \cos k(\theta_b+\theta) \qquad (5.9)$$

where

$$\alpha_{sk} = \int_{-b}^{b} \int_{-a}^{a} n(x_1 + x_2) T_s(x_1/a) T_k(x_2/b) p(x_1) p(x_2) \, dx_1 dx_2 \quad . \qquad (5.10)$$

The evaluation of this double integral for α_{sk} can be done by the two stages

$$n(b,\gamma)_k = \int_{-b}^{b} n(x_2+\gamma) T_k(x_2/b) p(x_2) \, dx_2 \qquad (5.11)$$

and

$$\alpha_{sk} = \int_{-a}^{a} n(b,x_1)_k T_s(x_1/a) p(x_1) \, dx_1 \qquad (5.12)$$

or vice versa.

When the two sinusoids are incommensurate, that is $\theta_a/\theta_b \neq m/n$ where m and n are integers, the only output at the frequency ω_a, where $\theta_a = \omega_a t$, is seen from eq. (5.9) to be for s=1 and k=0. On the other hand, if $\theta_a/\theta_b = m/n$ then other values of s and k may yield an output term at the frequency ω_a. The above approach carries over logically to any number of input sinusoids but the evaluation of the required integrals becomes difficult but for a few simply defined characteristics. To illustrate several aspects regarding the above results we consider the specific nonlinear characteristic $n(x) = x - (x^3/6)$. Firstly for a single sinusoidal input $x_1 = a \cos \theta_a$, the output harmonics are given by

$$a_s = 2 \int_{-a}^{a} \{x_1 - (x_1^3/6)\} T_s(x_1/a) p(x_1)\, dx_1 \quad (5.13)$$

with $b_s = 0$ as the characteristic is single valued. Also the nonlinearity has odd symmetry so that $a_s = 0$ for s even and in addition it is easily shown [3], using the orthogonal property of the Chebyshev polynomials $T_s(x)$, that since the highest power in n(x) is x^3 no harmonic greater than three is possible in the output. Substituting for $T_s(x/a)$ in eq. (5.13) and using eq. (4.31) yields

$$a_1 = a - (a^3/8) \quad (5.14)$$

and

$$a_3 = -a^3/24. \quad (5.15)$$

Supposing now the additional input $x_2 = b \cos (\theta_b + \theta)$ is added where ω_a and ω_b are incommensurate frequencies. The output magnitudes at the same frequencies as the terms of eqs. (5.14) and (5.15), that is at ω_a and $3\omega_a$, are from eq. (5.9) $2\alpha_{10}$ and $2\alpha_{30}$ respectively. Evaluating these coefficients using eqs. (5.11) and (5.12) we first obtain the zero modified nonlinearity (i.e. k=0) from

$$n(b,\gamma)_0 = \int_{-b}^{b} \{(x_2+\gamma) - (x_2+\gamma)^3/6\} p(x_2)\, dx_2$$

which gives

$$n(b,\gamma)_0 = \{1 - (b^2/4)\}\gamma - (\gamma^3/6). \tag{5.16}$$

Using this nonlinearity in eq. (5.12), which gives the same result as applying the sinusoid a to the nonlinearity $\{1 -(b^2/4)\}x-(x^3/6)$ we can write down $2\alpha_{10}$ and $2\alpha_{30}$ directly from eqs. (5.14) and (5.15), that is

$$2\alpha_{10} = \{1 - (b^2/4)\}a - (a^3/8) \tag{5.17}$$

and

$$2\alpha_{30} = -a^3/24 . \tag{5.18}$$

Finally, let us assume that $\omega_a/\omega_b=1/3$, so that the signal x_2 is the third harmonic of x_1. The term involving α_{10} in eq. (5.9) will still give an output at frequency ω_a but so also will a term with coefficient α_{21}, since $\cos 2\theta_a \cos (3\theta_a+\theta)$ is equal to $0.5 \{\cos (\theta_a+\theta) + \cos (5\theta_a + \theta)\}$. To evaluate α_{21} we first use eq. (5.11) with k=1 to give

$$n(b,\gamma)_1 = (b/2)\{1 - (b^2/8)\} - b\gamma^2/4 . \tag{5.19}$$

Using this expression in eq. (5.12) gives

$$\alpha_{21} = -a^2b/16 \tag{5.20}$$

and the total output at frequency ω_a is

$$2\alpha_{10} \cos \theta_a + 2\alpha_{21} \cos (\theta_a + \theta) .$$

Several points regarding these results should be noted. Since the nonlinearity has a soft spring characteristic, that is its slope, $n'(x)$, decreases monotonically as the input increases, the addition of the unrelated input, b, reduces the gain to a as $2\alpha_{10} = a_1 -(ab^2/4)$. This situation is true in general for a soft spring characteristic with the reverse case for a hard spring characteristic. When the input, b, is the third harmonic of a the output at frequency ω_a may increase or decrease, depending on the magnitude of b and its phase relative to a.

The most important point, however, is that the output at frequency, ω_a, is not now in phase with the input. It can be shown that the phase shift will have a maximum value of $\sin^{-1}(\alpha_{21}/\alpha_{10})$ when $\phi=\cos^{-1}(\alpha_{21}/\alpha_{10})$. For this choice of ϕ the magnitude of the output at the frequency, ω_a , is less than a_1.

Although we have considered a simple example, a characteristic involving a single cubic nonlinear term with an input consisting of a fundamental and third harmonic only, the results obtained are indicative of the general situation. The presence of harmonics in the nonlinearity input causes a phase shift to the fundamental and for a soft spring nonlinearity, if the phase of the harmonics is such as to produce near maximum phase shift to the fundamental, it usually undergoes a reduction in gain, as it would if the additional inputs were unrelated.

5.3 DF ACCURACY

When a limit cycle is excited in a feedback loop such that the nonlinearity input waveform differs greatly from a sinusoid the DF method obviously cannot predict the limit cycle accurately. For the feedback loop of Fig. 1.1 such a situation can occur when the filtering action of G(s) is such that its gain to one or more of the harmonics of the limit cycle frequency is not significantly less than its gain to the fundamental. In addition for a given G(s), assuming a limit cycle occurs, the DF prediction may be expected to be less accurate the greater the distortion produced by the nonlinearity. The DF accuracy problem is therefore a question of how to account, with confidence, for the harmonic terms in the input of the nonlinearity; also the procedure should not yield conservative results.

Two forms of incorrect results may be predicted by the DF. The case of most concern is when the method predicts stability, because no solution exists to eq. (4.48), yet the system is unstable and the second situation is when eq. (4.48) has a solution but no limit cycle exists. Using the terminology of reference [4] we will speak of these two cases as failures of the first and second type respectively.

There have been many investigations of DF accuracy [5-10] but these methods are difficult to apply, sometimes heuristic and often yield pessimistic results. Comparing the DF method for a SVNL with the circle criterion of Section 3.5 it will be noted that the DF lies within the real axis diameter of the circle so that the "uncertainty" surrounding the DF is primarily the "phase shifted" area of the circle. This can be reduced to one side of the DF for a monotonic nonlinearity by use of the off axis circle criterion, if the centre of the circle can be placed off the real axis to the other side of the DF. Since the circle criterion has to guarantee stability for any nonlinearity within a particular sector, one would expect for a smooth characteristic, which does not produce significant distortion, that the region around the DF for absolute stability could be reduced. The area will presumably be a function of both n(x) and G(s) but remains an unsolved and challenging problem. Below we consider two heuristic engineering approaches to estimating DF accuracy and also an exact procedure.

5.3.1 Distortion criteria

When the DF method indicates a limit cycle solution a good guide to the accuracy of the method is obtained by considering the loop to be opened at the nonlinearity input, injecting a signal corresponding to the solution and calculating the distortion in the signal fed back to the nonlinearity input. If for a symmetrical odd oscillation the percentage distortion, defined as the rms value of the harmonics to the rms value of the fundamental, is less than 5%, the amplitude and frequency of the DF solution are normally within 5% of the fundamental amplitude and frequency of the limit cycle. The error in the frequency is usually smaller than that in the amplitude [5]. For distortions greater than 10% the DF method cannot usually be relied upon.

The closeness of the $-1/N(a)$ and $G(j\omega)$ loci are of concern when the DF method predicts no limit cycle solution. In this case we can assume a solution which corresponds to a gain ratio of 0 db. and the smallest phase angle (see Section 4.6) to calculate the fed back distortion. A rough guide is that the smallest phase angle should be greater than the percentage distortion for the prediction of no limit cycle to be reliable. Again, if the distortion is greater than 10% the DF approach is not very appropriate [11].

Mees [10] has suggested an alternative approach of banding the $G(j\omega)$ locus with a width either side of magnitude $\max_{k>1} |G(jk\omega)|$ and requiring the banded locus to avoid the DF for stability. This technique appears to be too pessimistic in many cases and a more realistic width for the band might be $\max_{k>1} (a_k/a_1)$ $|G(jk\omega)|$ where the harmonic ratio a_k/a_1 takes its maximum value for all sinusoid magnitudes, a.

5.3.2 An exact banding method

A procedure for banding the frequency response locus of the linear elements which allows rigorous mathematical results to be obtained using the DF of the nonlinearity has been given by Mees [9]. The method creates a banded locus by drawing error circles of radius $\sigma(\omega)$ around the inverse Nyquist locus $\hat{G}(j\omega)$ at each frequency, ω. The following conclusions then apply:

(1) If the banded locus does not intersect $-N(a)$ then no odd symmetric limit cycle can exist

(2) There can be no limit cycle of frequency ω_o and amplitude, a_o, unless the error circle $\sigma(\omega_o)$ encloses the value $-N(a_o)$

(3) When the $-N(a)$ locus passes completely through the banded locus then there is at least one odd symmetric limit cycle with frequency, ω_o, and amplitude a_o for a $-N(a_o)$ contained within an error circle $\sigma(\omega_o)$.

The radii $\sigma(\omega)$ of the error circles are computed as follows. First, the disk $D(k_1,k_2)$ of the circle criterion (see Section 3.5) is drawn on the inverse Nyquist diagram. For this case the circle has its centre on the real axis and passes through the points $(-k_1,0)$ and $(-k_2,0)$ so that its radius, R, is equal to $(k_2-k_1)/2$. The shortest distance, $\rho(\omega)$, from a point P_k on the $G(j\omega)$ locus to the centre of the disc $D(k_1,k_2)$ is determined, where P_k lies outside the disc and is at a frequency $k\omega$, $k\neq 1$ and k odd.

The radius $\sigma(\omega)$ of the error circle around the point P_1 at frequency, ω, is then given by

$$\sigma(\omega) = R^2/(\rho(\omega) - R). \qquad (5.21)$$

5.4 STABILITY CONJECTURES

Two simple linearization conjectures have been give by Aizerman and Kalman regarding the stability of the system of Fig. 1.1. Although these conjectures are not valid in general both are true when certain constraints are placed on the frequency response, $G(j\omega)$, of the linear elements. Any counterexample to the conjectures is often in addition a type one failure of the DF method. Several of these cases are discussed in Section 5.5.

5.4.1 Aizerman Conjecture

Aizerman conjectured that if the system of Fig. 1.1 was stable with the nonlinearity replaced by a linear gain, k, for $k_1 < k < k_2$ then the nonlinear system would by asymptotically stable for a SVNL $n(x) \in (k_1, k_2)$. This corresponds to taking the real axis diameter of the disc $D(k_1, k_2)$ as the region to be avoided by $G(j\omega)$ for a stable system.

All the known counterexamples to the Aizerman conjecture, except one, are systems which possess limit cycles. The unique counterexample, due to Krasovskii [12], has

$$G(s) = -(s+1)/(s^2+s+1) \quad \text{and}$$

$$n(x) = \begin{cases} 0.9011x & |x| < 1 \\ [1 - \{e^{-2|x|}/|x|(1 + e^{-|x|})\}]x & |x| > 1 . \end{cases}$$

The Hurwitz sector of G(s) is $(-\infty, 1)$ and although $n(x) \in (0.901, 1)$ for some initial conditions the response becomes unbounded. This is a questionable counterexample since with unit loop gain the linear system has two poles at the origin and the gain of the nonlinearity tends asymptotically to unity for large inputs. If the Nyquist locus for $G(j\omega)$ and the diameter of the disc $D(k_1, k_2)$ are drawn they meet at

(-1, j0) where the amplitude is infinite and the frequency zero. Simulations have shown that if the intersection is avoided by restricting the nonlinearity such that $n(x) \in (\alpha, 1-\varepsilon)$ for several choices of α in the range $-\infty < \alpha < 1-\varepsilon$ and with ε small and positive, the system is stable [11].

No other counterexamples are known for second order systems. Bergen et al [13, 14] have investigated conditions required of third order transfer functions G(s) for the Aizerman conjecture to be valid. Since the required conditions for the coefficients of the terms in G(s) were obtained from a Lur'e type Liapunov function [13] and the Popov criterion [14], their only value is that one does not need to plot the Popov locus to prove stability for these specific cases, one of which is for a G(s) with no zeros and three real poles.

5.4.2 Kalman conjecture

Kalman conjectured that if the system of Fig. 1.1 were stable for $m_1 < k < m_2$ with a linear gain, k, replacing the nonlinearity then the nonlinear system would be stable for a SVNL $n(x) \in \{M\}$ with $n'(x) \in (m_1, m_2)$. Since a nonlinearity in the sector (k_1, k_2) will have $m_1 \leq k_1 < k_2 \leq m_2$, the line for $G(j\omega)$ to avoid on the Nyquist diagram to satisfy the Kalman conjecture is on the real axis from $-m_1^{-1}$ to $-m_2^{-1}$ and will include the diameter of $D(k_1, k_2)$. It is of interest to note that if bias inputs or disturbances are allowed to the system of Fig. 1.1, so that the nonlinearity input can contain a bias signal which in effect shifts its origin, then $m_1 \leq g(0,\gamma) \leq m_2$ for all γ, and this conjecture is equivalent to that of Aizerman for all possible bias signals to n(x).

The conjecture has been proved to be true using the off axis circle criterion for $n(x) \in \{M_m\}$ when $G(j\omega)$ satisfies the following constraints [15]:

(i) $G(j\omega)$ is a rational function of $j\omega$.

(ii) $|G(j\omega)|$ is bounded and monotonically decreasing for all $\omega \geq 0$.

(iii) The curvature of $G(j\omega)$ exists and is in the same non zero sense for all $\omega \geq 0$.

Several specific transfer functions are easily shown to have these properties, for example a G(s) with no zeros and all poles real and negative. For a given transfer function, not in the known set, the above three conditions can easily be checked computationally. Voronov [16], using the Popov criterion and relationships between the Popov and Nyquist loci, determined similar conditions for the Kalman conjecture to be true and also a more complete list of transfer function forms which satisfy the conditions. The Kalman conjecture does not seem realistic for other than $m_1 = k_1$ and $m_2 = k_2$, since when these conditions do not apply a system can be proved absolutely stable by the circle criterion yet violate the Kalman conjecture.

5.5 COUNTEREXAMPLES TO THE AIZERMAN CONJECTURE

One is tempted to search for common properties in transfer functions of systems which break down the Aizerman conjecture, but this is difficult since the poles of a transfer function and its frequency response can be changed by a pole transformation, which, of course, does not affect the stability of the system. We have seen, however, that the root locus if plotted for all $K \in [-\infty, \infty]$ is independent of a pole transformation. One characteristic which has been found common to the root loci of all known counterexamples is that they possess paths crossing the imaginary axis at a finite frequency and subsequently meeting on the real axis in the right hand side of the s plane [17,18]. Two specific counterexamples are described below.

Example 1

Fitts [19] considered a system with $G(s) = s^2/\{(s+0.01)^2 + 0.9^2\}\{(s+0.01)^2 + 1.1^2\}$ and a cubic nonlinearity $n(x)=x^3$. A portion of the frequency response and the root locus of this transfer function are shown in Figs. 5.1 and 5.2. At frequencies near to the resonant frequencies of the lightly damped poles the phase shift is almost 180°. Only a small harmonic content in a limit cycle is therefore sufficient to produce a small phase shift through the nonlinearity to the fundamental, thus providing a loop phase shift of 180°. Simulations by Fitts

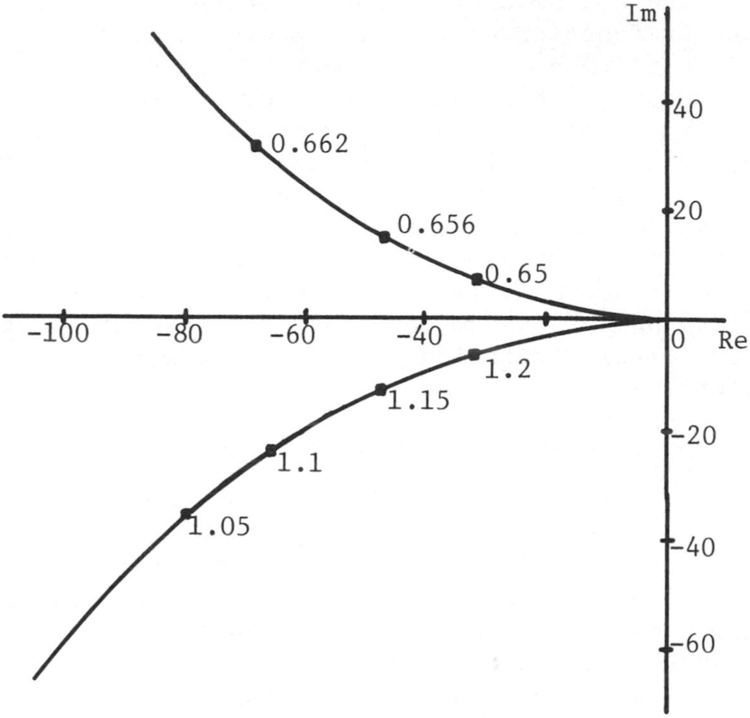

Figure 5.1 Frequency response for Example 1

revealed oscillations with a dominant higher harmonic of order dependent upon the initial conditions. Fig. 5.3 shows a limit cycle of frequency 0.61 rads/s. which has a dominant third harmonic giving a distortion of approximately 25%. This is obviously adequate to give the measured phase change of around 3° to the fundamental. Fitts examined the phase relationships of the harmonics in the limit cycle and showed that $\sum_{k=1}^{\infty} a_k V(jk\omega) = 0$ was a necessary condition for a limit cycle where a_k are positive coefficients and $V(j\omega) = \text{Im } G(j\omega)$. This can, of course, also be shown from the Popov criterion since it simply means that no limit cycle is possible in Fig. 1.1 when n(x) is single valued if $G(j\omega)$ lies only in the upper or lower half Nyquist plane.

Example 2

In reference [17] an example with $G(s)=-(s^2-0.5)(s+1)/s^2(s^2+3s+4)$ and n(x) a linear segmented character-

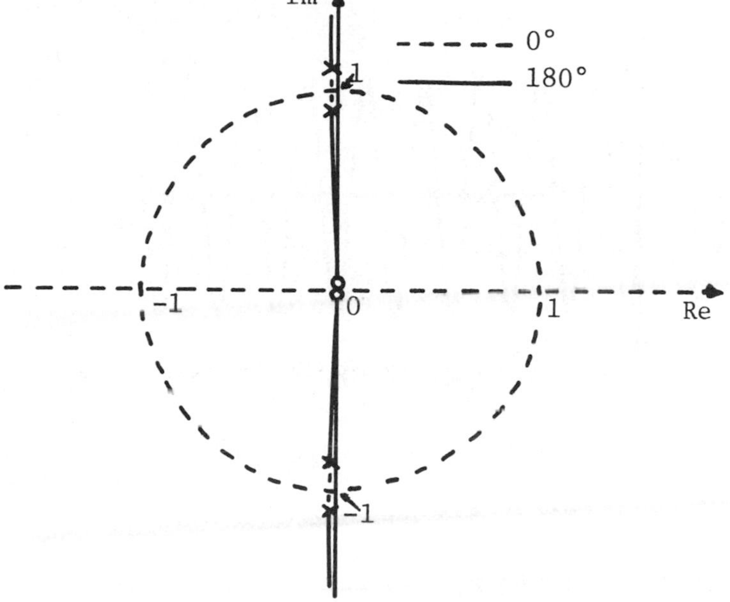

Figure 5.2 Root locus for Example 1

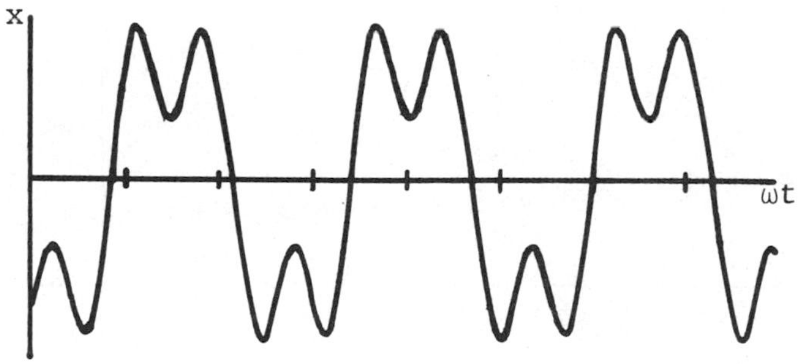

Figure 5.3 Limit cycle in Example 1

istic with initial slope 1.9, final slope 0.4 and a slope of −42 joining these lines was considered. The Hurwitz sector for the system is (0,2) but simulations gave the limit cycle at 0.88 rads/s. shown in Fig. 5.4. The phase shift through the nonlinearity

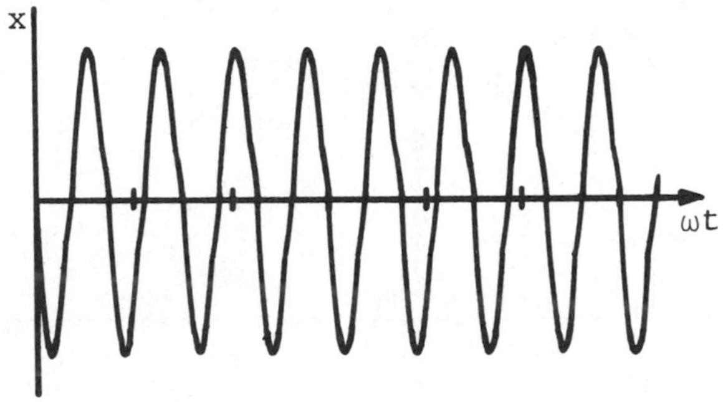

Figure 5.4 Limit cycle in Example 2

to the fundamental of the limit cycle was found to be 3.5°. The frequency response and root locus of -G(s) are shown in Figs. 5.5 and 5.6. Investigations of pole shifted versions of the system showed, as expected, significant changes in the limit cycle waveform at the input to the nonlinearity. For ρ=2 the distortion was greater than 30% and the phase shift to the fundamental through the nonlinearity around 35°.

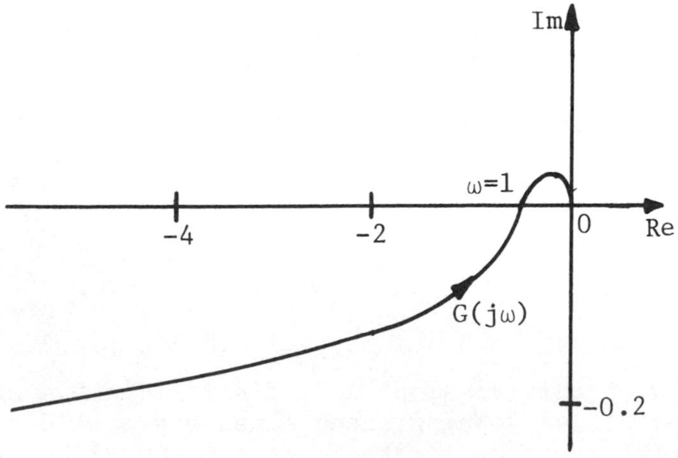

Figure 5.5 Frequency response for Example 2

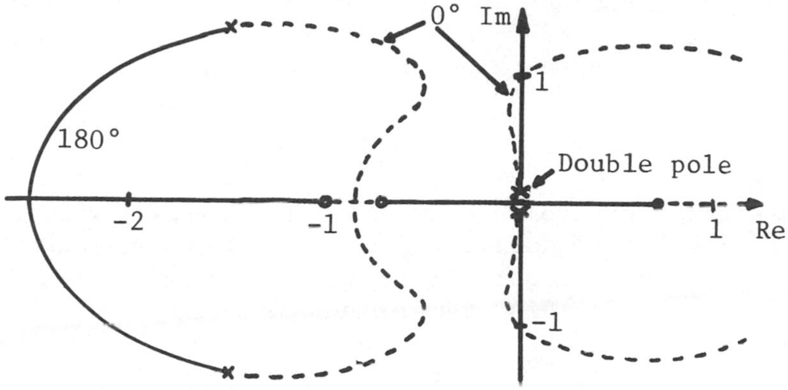

Figure 5.6 Root locus for Example 2

5.6 DF APPLICATIONS

In this section a few unusual examples are discussed to illustrate how the DF method can be used to determine the stability properties of some nonlinear systems. An example illustrating type 2 failure of the DF [4] is also given.

Example 3

Consider a system with $G(s)=K(s-1)/\{(s+1)(s^2+2s+2)\}$ and an ideal saturation characteristic, $n(x)$, of linear gain 4 which saturates when the input exceeds unity. The Hurwitz sector of $G(s)$ is $(-2.5, 2)$ so that from Vogt's theorem (see Section 3.7) the response must remain bounded as the final slope of the saturation characteristic is zero. Fig. 5.7 shows the root locus of $G(s)$ for both positive and negative values of K. With K=1 the null solution is unstable, since the saturation has a linear gain of 4, due to the real axis pole. We note, however, that $G(0)=-1/2$ and $n(x)$ has a bias gain of 2 for an input of ± 2. These points on the nonlinearity, which are where the Hurwitz sector bound line of slope 2 cuts the nonlinearity, are therefore possible equilibrium points. They are locally stable and simulations show that all initial condition responses converge to one of these equilibrium points.

On the other hand if we consider K=-1 the roots which move into the right hand side of the s plane as the gain increases on the root locus are now complex. The system therefore has a limit cycle the parameters of which can be obtained to better than 5% using the DF method.

Example 4

In digital control systems quantization, as well as sampling, of signals takes place. This can lead to difficulties in controlling unstable processes as illustrated in this example, where we neglect the effects of sampling. Consider Fig. 1.1 with a plant transfer function $G(s)=(s+4)/\{(s-0.5)(s^2+2s+2)\}$ and n(x) a quantization characteristic with output levels nh occuring at input magnitudes of (2n-1)h for n=1, 2...∞. The Hurwitz sector of G(s) is (0.25, 1.0) and the average gain of the quantizer is 0.5 which lies within the sector. However, the quantizer gain is zero for inputs less than h and the system is unstable due to the right hand side s plane pole of G(s).

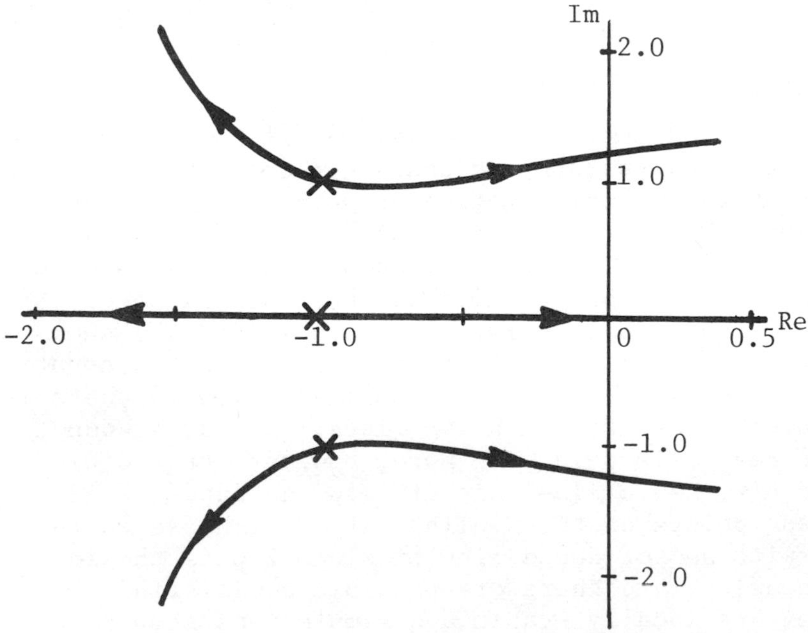

Figure 5.7 Root locus for Example 3

Since the maximum gain for stability is unity, which corresponds to complex poles moving into the right hand side of the s plane, an asymmetrical limit cycle builds up about the quantisation input level of +h or -h. The oscillation can in principle be evaluated using the SBDF but due to the large asymmetry in the waveform it does not give accurate results. An exact solution can be found using the method given in the next chapter. A simulation of the system gave an oscillation frequency of 1.21 rad/s, with the waveform at +1(or -1) for only 16% of the time. This compares with the DF method frequency of 1.41 rad/s.

Example 5

Rapp and Mees considered a system with $n(x)=-1/(1+x)$ and $G(s)=K/(s+a)^n$ for large values of n in the transfer function. Using a SBDF analysis unstable limit cycles were predicted for values of n greater than 8. An exact theoretical result due to Allwright [4] shows that no limit cycles can exist, thus the DF prediction must be incorrect. The predicted limit cycles were asymmetrical with a fundamental frequency, ω_1, where ω_1 is the minimum value of ω for $\underline{/G(j\omega_1)} = 180°$. For values of $n \geq 65$ more than one limit cycle solution was predicted by the SBDF method. It was argued that the DF predictions were inaccurate due to the poor filtering of the second harmonic provided by G(s). The ratio $|G(jk\omega_1)|/|G(j\omega_1)|$ was plotted as a function of n and was found to increase monotonically for $n > 4$ reaching a magnitude of 0.28 for n=10 and 0.60 for n=30.

5.7 DFs FOR FREQUENCY DEPENDENT NONLINEARITIES

An advantage of the DF method is that it can in principle be used to investigate limit cycles in more complex nonlinear systems than that of Fig. 1.1. For a system containing several nonlinear and linear elements, either in a single feedback loop or in multiple loops, the DF approach requires the assumption of one or more sinusoidal signals, possibly plus bias, at the input to each nonlinear element. When one has a good physical understanding of the problem of interest, or some knowledge has been gained from simulations, the appropriate assumptions are often readily apparent. For limit cycles in control systems the assumption that the input to each nonlinearity, $n_i(x)$, is a single sinusoid of frequency, ω, and of amplitude, a_i, and phase, ϕ_i, dependent on i will often be satisfactory. On the other hand, if one is investigating the effect of coupling two nonlinear oscillators, signals with two frequency components would normally have to be considered. In the case of nonlinear control systems the major difficulty is often that adequate low pass filtering may not exist between each nonlinear element, so that when a limit cycle exists its waveform is non sinusoidal at the input to each nonlinear element. DF results obtained for such situations will obviously be inaccurate or worse completely incorrect. In some cases it may be very appropriate to assume a fundamental plus third harmonic input to one or more nonlinear elements as discussed in the next section for the system of Fig. 1.1.

Some general observations for specific situations would seem appropriate. Nonlinearities connected directly in parallel can be represented by the sum of their individual DFs, but the DF of nonlinearities in series is not the product of their DFs. To obtain the DF for this latter case the DF of the nonlinearity corresponding to the series combination has to be obtained. In many systems two nonlinear elements occur separated by a linear transfer function which is not low pass. The DF of the combination may then be calculated and will be dependent on both the amplitude and frequency of the input. This can be a tedious process but in many instances knowledge of the system may indicate the need to evaluate the DF over only a small amplitude or frequency range. If both nonlinear elements are single valued the DF of the combination, due to the intervening linear dynamics, will be complex. In some instances, it may be appropriate to represent the nonlinearity by a function of the input and its derivative, that is $n(x,\dot{x})$ [3], and obtain the DF in the form $K_1x+K_2\dot{x}$, which in general will be amplitude and frequency dependent. The equations which yield a DF solution for some specific configurations of linear and nonlinear elements are given in references [3,20] although the reader should find these relationships easy to write down on the basis of the comments made earlier in the section.

5.8 DF EXTENSIONS

Various extensions of the DF method have been suggested to evaluate limit cycles more accurately. The simplest to apply is the refined DF. When a DF solution exists the fundamental and next lowest harmonic, or dominant harmonic, in the waveform fed back to the nonlinearity, assuming the loop to be opened in the manner used to calculate the distortion, are calculated. The refined DF, which it is claimed can evaluate the limit cycle more accurately, is then taken as the complex ratio of the fundamental output to the fundamental input for the previously calculated nonlinearity input.

An obvious extension to the DF method is to start by assuming the nonlinearity input to be a sinusoid together with its harmonics. For symmetrical odd system behaviour only odd harmonics will need to be included in the nonlinearity input. The difficulty with this procedure is that each harmonic term at the nonlinearity output will, in general, depend on the magnitude and phase relationship of all the input terms. The computations involved therefore become quite excessive if more than a few harmonics are considered in the nonlinearity input. Sufficient equations to solve for all the unknowns are obtained from the requirement that for Fig. 1.1 the loop gain for each harmonic must be minus one. In many instances a large improvement in accuracy may be obtained if only one harmonic is considered, especially the second harmonic when asymmetrical oscillations occur. Mees [21] has given a mathematical formalism for this procedure where he describes the transmission of the harmonics through the nonlinearity by a DF matrix. Evaluation of the terms in this matrix, which is related to the evaluation of coefficients similar to α_{sk} in eq. (5.10) (for the fundamental plus one harmonic), is possible analytically for simple power law or harmonic nonlinearities but must be done computationally for other characteristics.

To illustrate the procedure we consider a simple example by taking $n(x)=x-(x^3/6)$ and $G(s)=2(1-s)/s(1+s)$ in Fig. 1.1. $\underline{/G(j\omega)} = 180°$ for $\omega=1$ and the corresponding gain is 2.0. However, $G(j\omega)$ is not a good low pass filter, due to the all pass transfer function $(1-s)/(1+s)$, so that the DF solution may be appreciably in error. Since $N(a)=1-a^2/8$ the DF solution for a is given by $1-a^2/8=1/2$, i.e. $a=2$. To determine the limit cycle more accurately a nonlinear input x of the form

$$x = a \cos \theta_a + b \cos (3\theta_a + \phi)$$

may be assumed. Substituting in n(x), expanding and collecting terms gives for the output at frequency ω_a

$$\{a-(a/8)(a^2+2b^2+ab \cos \phi)\} \cos \theta_a + (1/8)a^2 b \sin \phi \sin \theta_a,$$

and at frequency $3\omega_a$

$$\{b-(b/8)(b^2+2a^2)-(a^3/24) \cos \phi\} \cos (3\theta_a+\phi) - (a^3/24) \sin \phi \sin (3\theta_a+\phi),$$

These terms can also be obtained using eqs. (5.9) and (5.10). Therefore

$$M_a(a,b) = [\{1 - (1/8)(a^2 + 2b^2 + ab \cos \phi)\}^2 + \{(1/8)ab \sin \phi\}^2]^{1/2}, \qquad (5.22)$$

$$M_b(a,b) = (1/b)[\{b - (b/8)(b^2 + 2a^2) + (a^3/24) \cos \phi\}^2 + \{(a^3/24) \sin \phi\}^2]^{1/2}, \qquad (5.23)$$

$$\psi_a(a,b) = \tan^{-1} \{-ab \sin \phi/(8 - a^2 - 2b^2 - ab \cos \phi)\}, \qquad (5.24)$$

$$\Psi_b(a,b) = \tan^{-1} \{a^3 \sin \phi/(24b - 3b^3 - 6a^2 b - a^3 \cos \phi)\}. \qquad (5.25)$$

where M_a and ψ_a are the nonlinearity gain and phase shift to the input a.

The loop equations for the limit cycle are

$$M_a(a,b)g(\omega) = 1 , \tag{5.26}$$

$$M_b(a,b)g(3\omega) = 1 , \tag{5.27}$$

$$\psi_a(a,b) + \theta(\omega) = (2j + 1)\ 180°, \tag{5.28}$$

$$\psi_b(a,b) + \theta(3\omega) = (2k + 1)\ 180°, \tag{5.29}$$

where j and k are integers.

A computer solution of the nonlinear equations gives two sets of values namely a=2.19, b=0.298, ϕ=138°, ω=0.882 and a=1.88, b=0.166, ϕ=-52°, ω=1.06. The former solution can be shown to be the one for the stable limit cycle. The waveform x(t) at the nonlinearity input computed from the above values of a,b,ϕ and ω is shown dotted in Fig. 5.8 together with that obtained by analog simulation. The agreement is seen to be much better than the DF solution. The nonlinearity output waveform, which is seen to contain a considerable amount of distortion, is also shown in the figure.

5.9 COMBINED MODES

When the system of Fig. 1.1 contains a SVNL and $\underline{/G(j\omega)}$ = 180° is satisfied for more than one value of ω several possible forms of oscillation may exist dependent on the system parameters. For example, if $\underline{/G(j\omega)}$=180° for $\omega=\omega_1$ and ω_2, which in general will be incommensurate frequencies, then a limit cycle may exist at either of the frequencies ω_1 and ω_2 or a combined mode oscillation, that is an oscillation consisting predominantly of both frequencies ω_1 and ω_2 may occur. To investigate the latter possibility the describing function for two unrelated sinusoidal inputs, SSDF, to a nonlinearity must be evaluated. Further to assess the stability of any solution the incremental two sinusoidal input describing function, ISSDF, must be found.

Using the same notation as Section 5.2 and denoting the SSDF component gains by $N_a(a,b)$ and $N_b(a,b)$

respectively, then from eqs. (5.9) and (5.12)

$$N_a(a,b) = 2\alpha_{10}/a \quad \text{and} \tag{5.30}$$

$$N_b(a,b) = 2\alpha_{01}/b \tag{5.31}$$

To evaluate the ISSDF, $N_{i\gamma}(a,b)$, we require the value of the bias output when the input is taken as $\gamma + a \cos\theta_a + b\cos(\theta_b+\phi)$. It is easily shown that

$$N_{i\gamma}(a,b) = \lim_{\gamma\to 0} \alpha_{00}/\gamma \tag{5.32}$$

where $\alpha_{00} = \int_{-b}^{b}\int_{-a}^{a} n(x_1 + x_2 + \gamma)\, p(x_1)p(x_2)\, dx_1 dx_2$ (5.33)

or alternatively

$$\begin{aligned} N_{i\gamma}(a,b) &= N_a(a,b) + (a/2)N_a'(a,b) \\ &= N_b(a,b) + (b/2)N_b'(a,b) \end{aligned} \tag{5.34}$$

where N_a' denotes partial differentiation of N with respect to the subscript a. The amplitude, a_o and b_o, of the sinusoids at the input to the nonlinearity in a combined mode oscillation are given using the DF method by the solutions of

$$1 + N_a(a_o,b_o)G(j\omega_1) = 0 \tag{5.35}$$

$$1 + N_b(a_o,b_o)G(j\omega_2) = 0\,. \tag{5.36}$$

The roots of

$$1 + N_{i\gamma}(a_o,b_o)G(s) = 0 \tag{5.37}$$

must all lie in the left hand side of the s plane for the mode to be stable.

Due to the cross modulation frequencies as well as the harmonics produced at the nonlinearity output with a two sinusoidal input signal, DF solutions for combined modes are often less accurate than for the single input situation. In particular it should be remembered that one or more of the cross modulation frequencies $s\omega_a \pm k\omega_b$, $s\neq 0$, $k\neq 0$ at the nonlinearity output may have a lower frequency than the inputs.

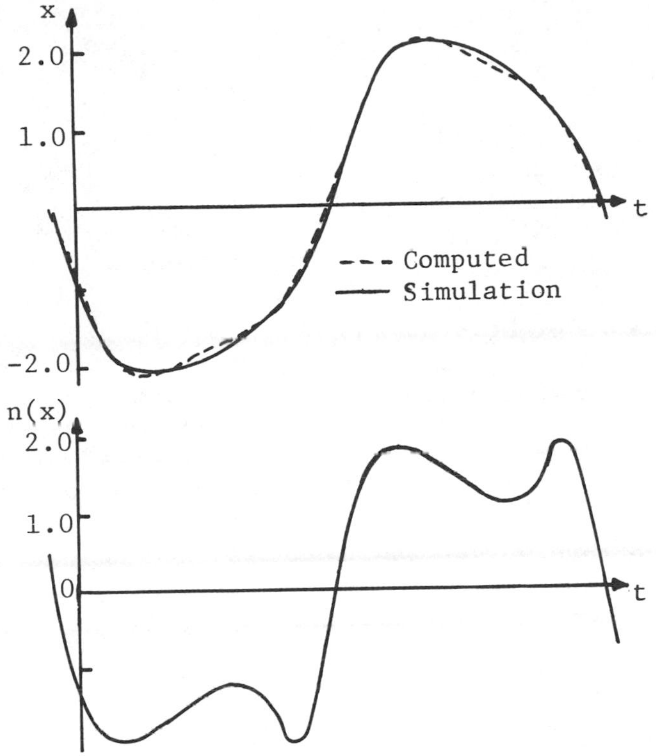

Figure 5.8 Limit cycle waveforms

When a solution is obtained for an oscillation the distortion technique discussed in Section 5.3 can be used to obtain an estimate of the accuracy of the combined mode solution.

To illustrate a combined mode oscillation we consider again the system mentioned in Section 4.11 which has an ideal relay and $G(s)=K(s+1)^2/s^3(s^2+0.3\lambda s+\lambda^2)$ [22]. For $\lambda=3.5$ the Nyquist locus with K=1 and root locus of G(s) are shown in Figs. 4.10 and 4.11 respectively. Denoting the two values of ω for which $\underline{/G(j\omega)}=180°$ by ω_1 and ω_2, with $\omega_1 > \omega_2$, we require for a single frequency limit cycle

$$g(\omega_n)N(a_n) = 1, \quad n=1, 2 \tag{5.38}$$

where $g(\omega)=|G(j\omega)|$. But for the ideal relay

$N_{i\gamma}(a) = N(a)/2$ so that the limit cycle will be stable if

$$1 + \{N(a_n)/2\}g(\omega_n) = 0 . \tag{5.39}$$

Using eq. (5.38) and knowledge of the root loci it is easily shown that for eq. (5.39) to have roots in the left hand side s plane

$$N(a_1) < N_{i\gamma}(a_n) < N(a_2) . \tag{5.40}$$

This can never be true for the limit cycle of frequency ω_1 and amplitude a_1, which is therefore always an unstable limit cycle solution, but will be true for ω_2 provided $a_2/a_1 > 2$, which corresponds to $\alpha>2$ where $\alpha=g(\omega_1)/g(\omega_2)$. α can be changed by varying λ and simulation results for various combinations of λ and K were found to give good agreement with the theory. It can be shown from eqs. (5.35) to (5.37), with amplitudes $a=a_1$, and $b=a_2$, that for a stable combined oscillation

$$N_{a_1}(a_1,a_2) < N_{i\gamma}(a_1,a_2) < N_{a_2}(a_1,a_2) \tag{5.41}$$

For the ideal relay

$$N_{a_1}(a_1,a_2) = 8E(k)/a_1\pi^2 \tag{5.42}$$

and

$$N_{a_2}(a_1,a_2) = 8B(k_1)/a_1\pi^2 \tag{5.43}$$

where $k=a_2/a_1 < 1$ and K(k) and E(k) are the complete elliptic integrals

$$K(k) = \int_0^{\pi/2} (1 - k^2\sin^2\theta)^{-1/2}d\theta \tag{5.44}$$

$$E(k) = \int_0^{\pi/2} (1 - k^2\sin^2\theta)^{1/2}d\theta \tag{5.45}$$

and

$$B(k) = \{E(k) - (1 - k^2)K(k)\}/k^2. \tag{5.46}$$

Using these results in eq. (5.41) with $N_{i\gamma}(a_1,a_2)$ evaluated from eq. (5.34) gives $a_1/a_2 < 0.91$ for stability which corresponds to $\alpha < 2$. Simulation results confirmed the existence of the combined mode for $\alpha < 2$ with the DF solutions in error with respect to the amplitude and frequency of the combined oscillation components by 5 - 10%.

5.10 SUMMARY

This chapter has considered various aspects of the accuracy of the DF method and demonstrated that several procedures can be used to check the reliability of the method. The major advantage of the DF method is that it is not in principle restricted to the specific feedback loop of Fig. 1.1 but may be used in more complicated situations. Indeed for the investigation of many situations, for example combined mode oscillations, it is the only approach available apart from numerical solution of the differential equations or simulation. Further the DF method can be used to consider additional problems such as forced oscillations, subharmonic oscillations, the effects of random inputs and so on which are outside the range of this monograph. The interested reader is referred to references [3,23] for information on these topics.

REFERENCES

1. Aizerman, M.A.: On a problem relating to the global stability of dynamic systems, Uspehi Mat. Nauk 4, No. 4 (1949).

2. Kalman, R.E.: "Physical and mathematical mechanisms of instability in nonlinear automatic control systems", Trans. ASME, J. Basic Eng., Vol. 79, pp. 553-563, Apr. 1957.

3. Atherton, D.P.: "Nonlinear control engineering", Van Nostrand Reinhold, London, 1975, Chapter five.

4. Rapp, P.E. and Mees, A.I.: "Spurious predictions of limit cycles in a non-linear feedback system by the describing function method", Int. J. Control, Vol. 26, No. 6, pp. 821-829, 1977.

5. Johnson, E.C.: "Sinusoidal analysis of feedback control systems containing nonlinear elements", Trans. AIEE, Pt. II, Vol. 71, pp. 169-181, July 1952.

6. Bass, R.W.: "Mathematical legitimacy of equivalent linearization of describing functions", Proc. 1st IFAC Congress, Butterworth, London, pp. 895-905, 1961.

7. Bergen, A.R. and Franks, R.L.: "Justification of the describing function method", SIAM J. Contr., Vol. 9, pp. 568-589, Nov. 1971.

8. Kudrewicz, J.: "The describing function method and free vibrations in nonlinear systems", Int. J. Circuit Theory and Appl., Vol. 1, pp. 49-57, Mar. 1973.

9. Mees, A.I. and Bergen, A.R.: "Describing function revisited", IEEE Trans. Automat. Contr., Vol. AC-20, pp. 473-478, Aug. 1975.

10. Lighthill, M.J. and Mees, A.I.: "Stability of nonlinear feedback systems", Recent mathematical Developments in Control, D.J. Bell, Ed., Academic press, pp. 1-20, London, 1973.

11. Rao, M.U.: "Some aspects of the stability of nonlinear control systems", M.Sc. thesis, University of New Brunswick, 1975.

12. Narendra, K.S. and Taylor, H.H.: "Frequency domain criteria for absolute stability", Academic Press, New York, 1973.

13. Bergen, A.R. and Willems, I.J.: "Verification of Aizerman's conjecture for a class of third-order systems", IRE Trans. Automat. Contr., Vol. AC-7, No. 3, 1962.

14. Bergen, A.R. and Baker, R.A.: "On third-order systems and Aizerman's conjecture", IEEE Trans. Automat. Contr., Vol. AC-17, pp. 220-222, April 1972.

15. Fannin, D.R. and Rushing, A.J.: "Verification of the Kalman conjecture based on locus curvature", Proc. IEEE, Vol. 62, No. 4, pp. 542-543, April, 1974.

16. Voronov, A.A.: "Systems with differentiable non-decreasing non-linearity absolutely stable in the Hurwitz angle", Proc. 7th IFAC Congress, Helsinki, pp. 1731-1736, 1978.

17. Shankar, S.: "Stability analysis of nonlinear control systems", M.Sc. thesis, University of New Brunswick, 1976.

18. Atherton, D.P. and Shankar, S.: "Some observations on the Aizerman conjecture", Proc. IEE, Vol. 126, No. 12, pp. 1305-1306, 1979.

19. Fitts, R.E.: "Two counterexamples to Aizerman's conjecture", IEEE Trans. Automat. Contr., Vol. AC-11, pp. 553-556, July, 1966.

20. Davison, E.J. and Constantinescu, D.: "A describing function technique for multiple non-linearities in a single-loop feedback system", IEEE Trans. Automat. Contr., Vol. AC-16, pp. 56-60, Feb. 1971.

21. Mees, A.I.: "The describing function matrix", J. Inst. Math. Appl. vol. 10, pp. 49-67, 1972.

22. Choudhury, S.K. and Atherton, D.P.: "Limit cycles in high-order nonlinear systems", Proc. I.E.E., 121, pp. 717-724, 1974.

23. Gelb, A. and Vander Velde, W.E.: "Multiple-input describing functions and nonlinear system design", New York, McGraw-Hill, 1968.

PROBLEMS

1. For example 1, Section 5.5 evaluate the distortion in the fed back signal assuming an oscillation frequency of 0.61 rad/s.

2. Evaluate the DF for the asymmetrical nonlinearity

$$n(x) = \begin{cases} k_1 x & x > 0 \\ k_2 x & x < 0 \end{cases}$$

What are the conditions on $k_1 k_2$ for Fig. 1.1 to be stable with this nonlinearity and $G(s)=1/(s+1)(s+2)(s+3)$. Can you predict stability by the DF method for n(x) lying outside the Aizerman sector. How do your results compare with those for problem 9 of chapter 3.

3. Evaluate the SSDF for unrelated inputs to the nonlinearity $n(x)=x^n$ for n odd.

4. Evaluate the SSDF for unrelated inputs to the nonlinearity $n(x)=A \sin mx$.

5. Evaluate the limit cycle in a feedback loop with $n(x)=x-(x^3/6)$ and $G(s)=2e^{-2s}/s$ using (a) the DF method (b) the refined DF (c) the DF method assuming a fundamental plus third harmonic as the input to the nonlinearity.

6. The feedback loop of Fig. 1.1 has an ideal relay nonlinearity and a transfer function G(s). Evaluate the limit cycle if (a) $G(s)=30/(s-1)(s^2+2s+10)$ and (b) $G(s)=30/(s-3)(s^2+2s+10)$. Will the limit cycle be symmetrical in both cases?

7. Prove eq. (5.34) for the ISSDF. Evaluate the ISSDF for (a) $n(x)=x^3$ and (b) $n(x)=A \sin mx$.

8. Evaluate the expressions given in eqs. (5.42) and (5.43) for the SSDF of an ideal relay.

9. Determine the limit cycle using a fundamental plus third harmonic DF for the nonlinearity $n(x)=\sin x$ in the feedback loop of Fig. 1.1 with

$G(s)=3e^{-2s}/(s+1)$.

10. The system of problem 2 has a constant bias input of 1 unit. How does this input affect the behaviour of the system if (a) $k_1=18$ and $k_2=2$ and (b) $k_1=72$ and $k_2=1/2$. Determine any limit cycles.

11. Investigate the possible modes of oscillation in a system with $G(s)=s(s^2+1.96s+1.96)/s(s+1)^2(s^2+0.384s+2.56)$ and (a) an ideal relay nonlinearity and (b) a cubic.

12. Evaluate using the DF method the parameters of the limit cycle in example 3, section 5.6, for K=-1.

13. Evaluate the DF of

$$n(x) = \begin{cases} m_1 x & x < \delta \\ h & x < \delta \end{cases}$$

with $h > m\delta$, by considering the nonlinearity as a parallel combination of two nonlinear characteristics.

CHAPTER 6

Limit Cycles in Relay Systems

6.1 INTRODUCTION

Many control systems employ relays or use controllers that operate in a relay type mode. Perhaps the most well-known relay control system is the on-off temperature controller used in buildings or for ovens, whilst other applications include relay control of motors and the attitude control of satellites using gas jets. In addition, other forms of controllers, such as a pulse width modulator, PWM, and a pulse frequency modulator, PFM, have a behaviour similar to that of a relay, in the sense that the magnitude of the output can have only a few possible values with the width and/or spacing of the pulses controlled by the input signal. It is because of this unique nature of the output from a relay, when an oscillation exists in the basic feedback loop of Fig. 1.1, that an exact solution for the limit cycle is possible.

Fig. 4.3 shows the characteristic of a relay with dead zone and hysteresis. Three other relay characteristics which are often used may be obtained as special cases of the above characteristic. If $\Delta=0$ one has the relay with dead zone, if $\delta=0$ one has the on-off relay with hysteresis and finally if $\Delta=0$ and $\delta=0$ one has the on-off or ideal relay. When a limit cycle occurs in the system of Fig. 1.1, with the nonlinearity a relay with dead zone and hysteresis, the waveform at the relay output will

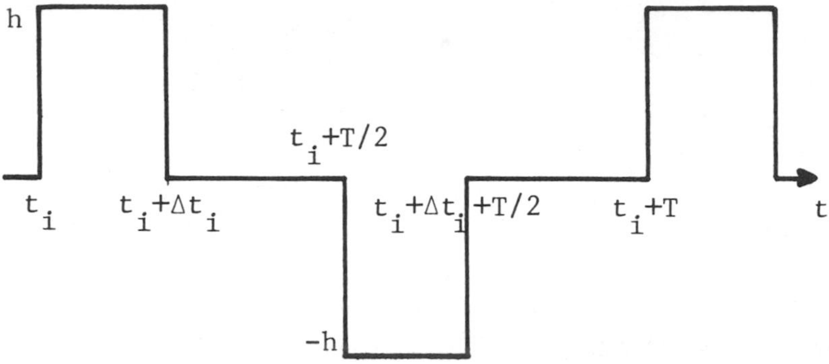

Figure 6.1 Relay output waveform

usually be of the form shown in Fig. 6.1. This waveform assuming the relay output magnitude is known, has only three unknown parameters, namely the switch on time t, the switch off time t+Δt and the period T. For the relay to switch at these time instants the input signal must have appropriate values, namely δ+Δ at time t and δ−Δ at time t+Δt. The limit cycle can thus be determined once the response of the plant transfer function G(s) to the assumed relay output waveform is found. The starting point in this analysis is therefore an assumed relay output waveform in contrast to the DF approach where the assumed waveform is that of the nonlinearity input.

In the next section we therefore consider the problem of evaluating the output of G(s) when the input, $y_i(t)$, is a periodic pulse waveform of the form shown in Fig. 6.2. This can be done directly in the time

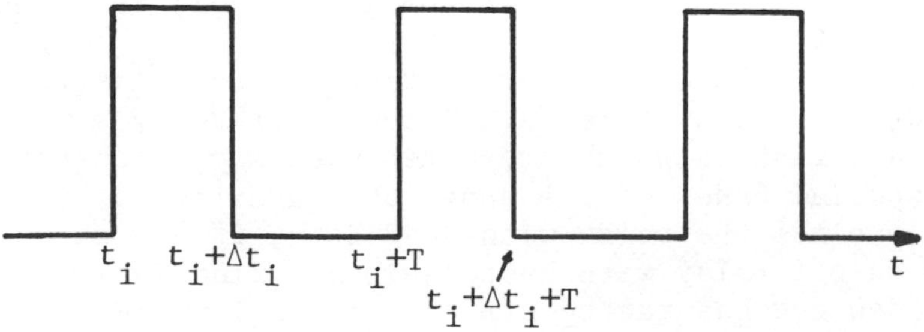

Figure 6.2 Periodic pulse waveform

domain [1,2,3] using a Laplace transform or state space approach, or using a frequency domain approach [4] by first expanding $y_i(t)$ as a Fourier series. Although the procedures are equivalent we will concentrate on the frequency domain approach here, primarily because of the ease of comparison with the DF method.

6.2 THEORETICAL BACKGROUND

Consider the periodic pulse train $y_i(t)$ shown in Fig. 6.2. It is straightforward to show that $y_i(t)$ can be expressed as the Fourier series

$$y_i(t) = (h\Delta t_i/T) + (h/\pi) \sum_{n=1}^{\infty} (1/n)\{\sin n\omega\Delta t_i \cos n\omega(t-t_i) + (1 - \cos n\omega\Delta t_i) \sin n\omega(t-t_i)\}. \quad (6.1)$$

If $y_i(t)$ is the input to a linear system with transfer function $G(s)$, then the Fourier series for the output $c_i(t)$, assuming $\lim_{s\to\infty} G(s)=0$ and $G(0)$ is finite, is given by

$$c_i(t) = (h\Delta t_i G(0)/T) + (h/\pi) \sum_{n=1}^{\infty} (g_n/n)\{\sin n\omega\Delta t_i \cos (n\omega t - n\omega t_i + \phi_n) + (1 - \cos n\omega\Delta t_i) \sin (n\omega t - n\omega t_i + \phi_n)\}. \quad (6.2)$$

where

$$G(jn\omega) = g_n e^{j\phi_n} = U_G(n\omega) + jV_G(n\omega) . \quad (6.3)$$

Differentiating eq. (6.2) the Fourier series for $\dot{c}_i(t)$, provided $\lim_{s\to\infty} sG(s)=0$, is

$$\dot{c}_i(t) = (\omega h/\pi) \sum_{n=1} g_n\{-\sin n\omega\Delta t_i \sin (n\omega t - n\omega t_i + \phi_n) + (1 - \cos n\omega\Delta t_i) \cos (n\omega t - n\omega t_i + \phi_n)\}. \quad (6.4)$$

When $\lim_{s\to\infty} sG(s) \neq 0$, c(t) will have discontinuities at the relay switching instants t_i and $t_i+\Delta t_i$. Since the Fourier series tends to the mean value of the function at a discontinuity, a term $0.5\{y_i(t^+) - y_i(t^-)\} \lim_{s\to\infty} sG(s)$ with appropriate sign must be included in eq. (6.4) to obtain the correct value of $\dot{c}_i(t)$ at these two time instants.

Let us now define the A locus [5] of a transfer function G(s) by

$$A_G(\theta,\omega) = \mathrm{Re}\, A_G(\theta,\omega) + j\, \mathrm{Im}\, A_G(\theta,\omega) \tag{6.5}$$

with

$$\mathrm{Re}\, A_G(\theta,\omega) = \sum_{n=1}^{\infty} V_G(n\omega) \sin n\theta + U_G(n\omega) \cos n\theta \tag{6.6}$$

and

$$\mathrm{Im}\, A_G(\theta,\omega) = \sum_{n=1}^{\infty} (1/n)\{V_G(n\omega) \cos n\theta - U_G(n\omega) \sin n\theta\}. \tag{6.7}$$

$A_G(\theta,\omega)$ is a generalized summed frequency locus with its real and imaginary values at a particular frequency ω depending upon weighted values, according to the choice of θ, of the real and imaginary values of $G(j\omega)$ at the frequencies $n\omega$ for $n=1, \ldots.\infty$. In particular for $\theta=0$ the real (imaginary) part of $A_G(0,\omega)$ depends only upon the real (imaginary) part of $G(j\omega)$ at frequencies $n\omega$. Using the above relationships in eqs. (6.2) and (6.4) it is easily shown that the expressions for $c_i(t)$ and $\dot{c}_i(t)$ can be written

$$c_i(t) = (hG(0)\Delta t_i/T) + (h/\pi)\{\mathrm{Im}\, A_G(-\omega t + \omega t_i,\omega) - \mathrm{Im}\, A_G(-\omega t + \omega t_i + \omega\Delta t_i,\omega) \tag{6.8}$$

and

$$\dot{c}_i(t) = (\omega h/\pi)\ \{\mathrm{Re}\, A_G(-\omega t + \omega t_i,\omega) - \mathrm{Re}\, A_G(-\omega t + \omega t_i + \omega\Delta t_i,\omega)\}. \tag{6.9}$$

Eq. (6.9) may also be seen to follow directly from eq. (6.8) since from the definitions of eq. (6.6) and (6.7)

$$\frac{d\{\text{Im } A_G(\theta,\omega)}{d\theta} = -\text{Re } A_G(\theta,\omega) \ . \tag{6.10}$$

In many control situations where limit cycles exist they have odd symmetry. For this case the input y(t) to G(s) can be written

$$y(t) = \sum_{i=1}^{2} y_i(t) \tag{6.11}$$

with

$$y_2(t) = -y_1(t + 0.5T) \ . \tag{6.12}$$

The resultant output c(t) will be the sum of the two responses $c_1(t)$ and $c_2(t)$, and can be shown to be given by

$$c(t) = (2h/\pi) \{\text{Im } A_G^{\circ}(-\omega t + \omega t_1,\omega) - \text{Im } A_G^{\circ}(-\omega t + \omega t_1 + \omega\Delta t_1,\omega)\} \tag{6.13}$$

and

$$\dot{c}(t) = (2\omega h/\pi) \{ \text{Re } A_G^{\circ}(-\omega t + \omega t_1,\omega) - \text{Re } A_G^{\circ}(-\omega t + \omega t_1 + \omega\Delta t_1,\omega)\} \tag{6.14}$$

where t_1 is the instant at which $y_1(t)$ switches positive and Δt_1 is the pulse width. In this case A° denotes the A locus with odd terms only, that is $n=1,3,5,\ldots\infty$ in eqs. (6.6) and (6.7).

If the signal $y_i(t)$ is an impulse train, so that $\Delta t_i \to 0$ and $h\Delta t_i \to M_i$, the impulse strength, then it is easily shown, since

$$\text{Im } A_G(\omega t + \omega\Delta t,\omega) = \text{Im } A_G(\omega t,\omega) - \omega\Delta t \text{ Re } A_G(\omega t,\omega) \tag{6.15}$$

that

$$c_i(t) = (M_i G(0)/T) + (M_i \omega/\pi)\text{Re } A_G(-\omega t + \omega t_i, \omega) \tag{6.16}$$

provided $\lim_{s\to\infty} sG(s)=0$. When the limit condition is not satisfied appropriate corrections have to be made, as mentioned earlier, at the instant of occurrence of the impulse. Finally if y(t) is an odd symmetrical impulse train then

$$c(t) = (2M_i \omega/\pi)\text{Re } A_G^\circ(-\omega t + \omega t_i, \omega) . \tag{6.17}$$

6.3 'A' LOCI FOR SPECIFIC TRANSFER FUNCTIONS

Firstly, it is appropriate to list some of the properties of the A loci defined by eqs. (6-5) - (6-7).

(i) The A locus for $\theta=0$ is identical with the Tsypkin locus [4], $\Lambda(\omega)$, if $\lim_{s\to\infty} sG(s)=0$, apart from a constant factor. The relationship is

$$\Lambda(\omega) = (4h/\pi)\ A(0,\omega). \tag{6.18}$$

The A locus can thus be regarded as a generalised Tsypkin locus.

(ii) The A locus satisfies the superposition property, that is, if the linear plant $G(s) = G_1(s) + G_2(s)$ then

$$A_G(\theta,\omega) = A_{G_1}(\theta,\omega) + A_{G_2}(\theta,\omega) . \tag{6.19}$$

(iii) If $G'(s) = G(s)e^{-sT_d}$ then

$$A_{G'}(\theta,\omega) = A_G(\theta + \omega T_d, \omega). \tag{6.20}$$

(iv) The loci are periodic in θ with period 2π, that is

$$A_G(\theta,\omega) = A_G(\theta + 2\pi, \omega). \tag{6.21}$$

(v) For A° the periodicity is odd, that is

$$A_G^\circ(\pi - \theta,\omega) = -A_G^\circ(-\theta,\omega). \tag{6.22}$$

Since any plant transfer function G(s) can be written in terms of a summation of transfer functions having a real or complex pair of poles, use of eq. (6.19) allows A loci for most transfer functions to be obtained in terms of the A loci of the few basic transfer functions given in Table 6.1. In addition use of eq. (6.20) enables A loci for transfer functions with time delays to be found.

TABLE 6.1

Basic Transfer Function Types

Type No.	Transfer Function	Type No.	Transfer Function
1	$1/s\tau$	6	$s\tau/(1+s\tau)^2$
2	$1/(1+s\tau)$	7	$1/s^3\tau^3$
3	$1/[s\tau(1+s\tau)]$	8	$1/(1+s\tau)^3$
4	$1/s^2\tau^2$	9	$1/(s^2+2\zeta\omega_0 s+\omega_o^2)$
5	$1/(1+s\tau)^2$	10	$s/(s^2+2\zeta\omega_0 s+\omega_o^2)$

When these basic transfer functions are substituted in eqs. (6.6) and (6.7) the resultant infinite series for both odd or all values of n can be evaluated in a closed form. The results are given in Appendix II, in Table II.1 for A° and Table II.2 for A, with $\lambda = \omega\tau$

6.4 LIMIT CYCLE SOLUTION

Having formulated a procedure for evaluating the output, c(t), of a linear transfer function to a periodic pulse input waveform we are now in a position to find limit cycle solutions for the relay system of Fig. 1.1.

6.4.1 Relay with no dead zone

We consider first the case where the relay has no dead zone, that is $\delta=0$. The relay switches between $\pm h$ so that for a symmetrical odd oscillation the pulse width $\omega\Delta t_1=\pi$. In addition, we may take the instant t_1, at which the relay switches positive , to be the zero time reference so that eqs. (6.13) and (6.14) give

$$c(t) = (4h/\pi)\operatorname{Im} A_G^\circ(-\omega t,\omega) \qquad (6.23)$$

and

$$\dot{c}(t) = (4\omega h/\pi)\operatorname{Re} A_G^\circ(-\omega t,\omega). \qquad (6.24)$$

For the relay to switch positive as assumed at $t=0$ the switching conditions

$$x(0^-) = \Delta \qquad (6.25)$$

and

$$\dot{x}(0^-) > 0 \qquad (6.26)$$

must be satisfied. Also with no external input

$$x(t) = -c(t). \qquad (6.27)$$

Further for the assumed output to be valid the input $x(t)$ must remain greater than $-\Delta$ for $0 < t < \pi/\omega$. This continuity condition will usually be satisfied by a solution, unless $G(s)$ has lightly damped poles, in which case it may be necessary to check its validity by computing $c(t)$ from eq. (6.23) for $0<t<\pi/\omega$. Therefore, from eqs. (6.23) - (6.27) satisfaction of the switching conditions requires

$$\operatorname{Im} A_G^\circ(0,\omega) = -\pi\Delta/4h \qquad (6.28)$$

and

$$\operatorname{Re} A_G^\circ(0,\omega) < 0 \qquad (6.29)$$

provided $\lim_{s\to\infty} sG(s) = 0$
For a given $G(s)$ eq. (6.28) can be solved for the

oscillation frequency ω and the condition (6.29) checked. It is interesting to note that the line with imaginary part $-\pi\Delta/4h$ and real part less than zero is the DF of the relay, the solution for a limit cycle is simply given by the intersection of $A_G^\circ(0,\omega)$ with the relay DF. One can therefore clearly see the accuracy of a DF solution if both $G(j\omega)$ and $A_G^\circ(0,\omega)$ are plotted.

As a specific example consider $G(s)=K/\{s(1+s\tau)\}$. For this transfer function use of the tables gives

$$\text{Im } A_G^\circ(0,\omega) \quad = \quad (K\pi\tau/4)\{(-\pi/2\lambda) + \tanh \ (\pi/2\lambda)\} \tag{6.30}$$

where $\lambda=\omega\tau$. Using eq. (6.28) the solution for the limit cycle frequency is given by

$$(\pi/2\lambda) - \tanh \ (\pi/2\lambda) \quad = \quad \Delta/hK\tau \tag{6.31}$$

since $\text{Re } A_G^\circ(0,\omega) < 0$ for all $\omega > 0$. The DF method for the same problem yields the equation

$$\lambda(1 + \lambda^2) \quad = \quad 4hK\tau/\pi\Delta \tag{6.32}$$

for the approximate frequency of oscillation.

Fig. 6.3 shows the graphical solution to the problem with $K=1$, $\tau=1$ and $h/\Delta=3$. For this case the exact limit cycle frequency is 1.365 rad/s. and that given by the DF method is 1.352 rad/s. If the parameter τ of the transfer function $G(s)$ is varied, then as τ is reduced the filtering action of $G(s)$ becomes worse and the DF solution can be expected to become less accurate. This is illustrated in Fig. 6.4 which shows the exact limit cycle frequency and the percentage error of the DF solution as a function of τ.

6.4.2 <u>Relay with dead zone</u>

To evaluate the parameters of a symmetrical odd limit cycle in the system of Fig. 1.1, when the nonlinearity is a relay with dead zone, one can again assume the positive switching time instant t_1 to be zero. In this case, however, the pulse width Δt_1 is unknown and its value together with the period T can be

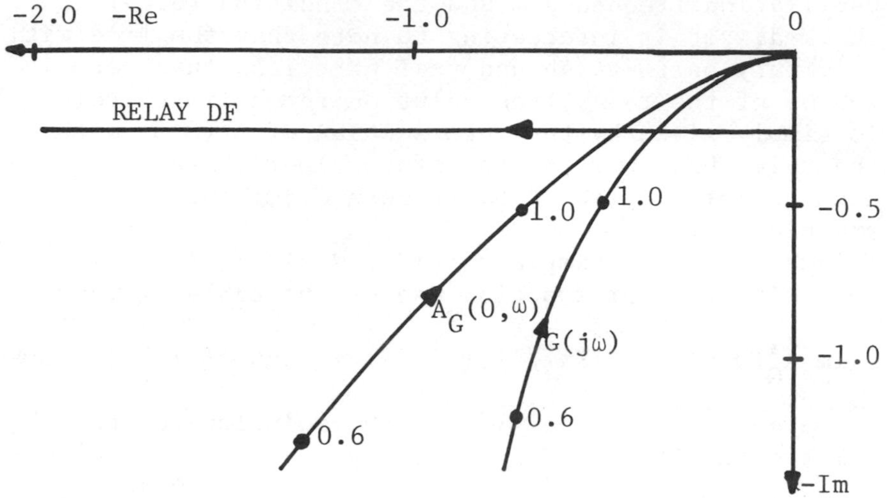

Figure 6.3 Solution by Tsypkin method

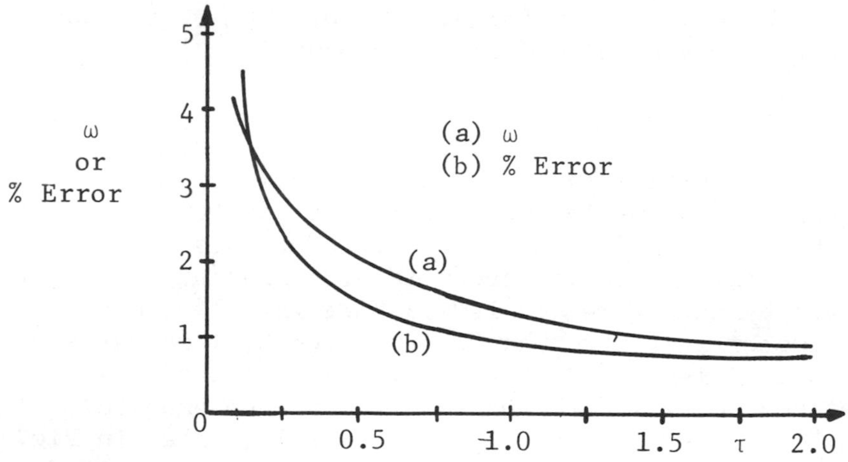

Figure 6.4 Comparison of predicted limit cycle frequencies

evaluated from the equations given by the switching conditions. The switching conditions are

$$x(0^-) = \delta + \Delta \tag{6.33}$$

and

$$\dot{x}(0^-) > 0 \tag{6.34}$$

for the positive switching and

$$x(\Delta t_1^-) = \delta - \Delta \tag{6.35}$$

$$\dot{x}(\Delta t_1^-) < 0 \tag{6.36}$$

for the switching from the positive value back to zero. Again for no input signal, $r(t)$, to the feedback loop $x(t)=-c(t)$ with $c(t)$ and $\dot{c}(t)$ given by eqs. (6.13) and (6.14) respectively with $t_1=0$. Using these facts in eqs. (6.33) - (6.36) gives

$$\text{Im}\{A_G^\circ(0,\omega)-A_G^\circ(\omega\Delta t_1,\omega)\} = -\pi(\delta + \Delta)/2h \tag{6.37}$$

$$\text{Re}\{A_G^\circ(0,\omega)-A_G^\circ(\omega\Delta t_1,\omega)\} < 0 \tag{6.38}$$

and

$$\text{Im}\{A_G^\circ(0,\omega)-A_G^\circ(-\omega\Delta t_1,\omega)\} = \pi(\delta - \Delta)/2h \tag{6.39}$$

$$\text{Re}\{A_G^\circ(0,\omega)-A_G^\circ(-\omega\Delta t_1,\omega)\} < 0 \,. \tag{6.40}$$

The nonlinear algebraic eqs. (6.37) and (6.39) can be solved for the unknown parameters Δt_1 and ω of the limit cycle. These equations require correction terms to be added to the right hand side when $\lim_{s\to\infty} G(s)\neq 0$. The easiest way to handle these cases of not strictly proper transfer functions is probably by evaluating $c(t)$ from the sum of $G(\infty)y(t)$ and the Fourier series obtained for the output of $G^*(s)$ with $y(t)$ as input, where $G(s)=G(\infty)+G^*(s)$ and $G^*(s)$ is strictly proper.

Inequalities (6.38) and (6.40) with the solution parameters substitued must be satisfied for the solution to be valid. When $\lim_{s\to\infty} sG^*(s)\neq 0$ a correction factor must be added to the right hand side of these inequalities. Since this factor will be positive for $\lim_{s\to\infty} sG^*(s)>0$ the above conditions are necessary but not sufficient. Once a solution is obtained the continuity conditions that

$$x(t) > \delta - \Delta \quad \text{for } 0 < t < \Delta t \,, \tag{6.41}$$

and

$$-(\delta + \Delta) < x(t) < \delta - \Delta \text{ for } \Delta t_1 < t < \pi/\omega \tag{6.42}$$

should strictly be checked; however, they will most probably be satisfied unless G(s) has lightly damped poles.

For specific examples it may be possible to solve eqs. (6.37) and (6.39) analytically when the transfer function G(s) has simple forms. In general, however, the equations will have to be solved numerically or graphically. To obtain a graphical solution the loci $A_G^\circ(0,\omega) - A_G^\circ(\omega\Delta t_1,\omega)$ and $A_G^\circ(0,\omega) - A_G^\circ(-\omega\Delta t_1,\omega)$ can be plotted in the complex plane as a function of the frequency ω for different choices of the angle $\theta=\omega\Delta t_1$. Two sets of relationships between ω and θ are obtained where these loci intersect the two straight lines parallel to the real axis and at distances from it of respectively $-\pi(\delta+\Delta)/2h$ and $\pi(\delta-\Delta)/2h$. The values of ω and θ which satisfy both these relationships can then be found by plotting the two sets of values in an ω - θ plane as shown in Fig. 6.5. By hand this is a laborious procedure but with modern computer graphics facilities the tedium can be removed. An advantage of the graphical approach, like that of the DF method, is that it allows one to more readily interpret the effect on the limit cycle solution of varying a system parameter.

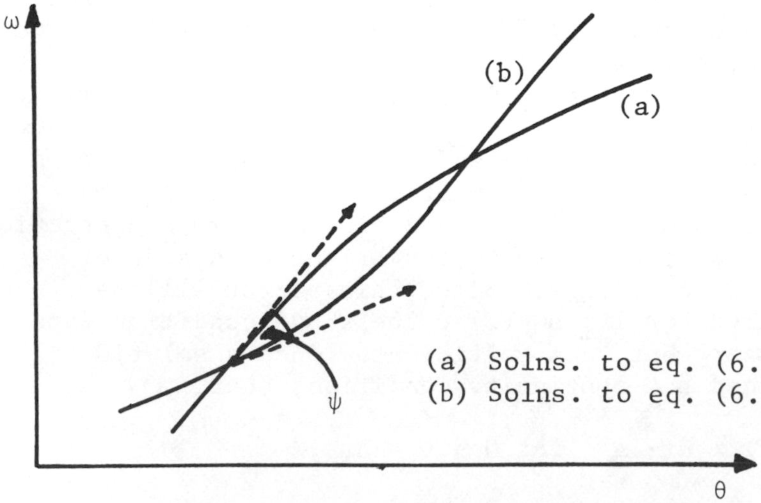

Figure 6.5 Solution from ω-θ plane

Since it is possible to store the algebraic expressions for the A loci, given in Appendix II, as functions in the computer the required graph can be quickly generated. These loci are shown in Fig. 6.6 for $G(s)=8/s(1+s)^2$.

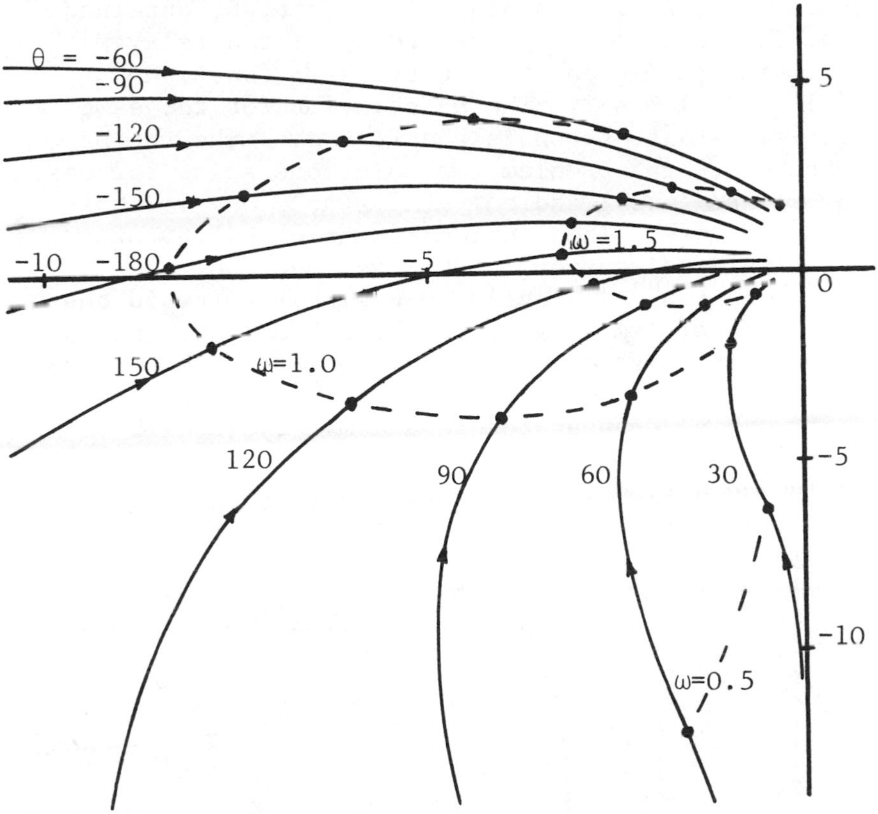

Figure 6.6 Graphs of A loci

It is also possible to obtain a graphical solution from a plot of the DF of the relay, which is a function of θ, and plots of a summed frequency locus, $G_m(\theta,\omega)$, drawn for various values of θ. Since this particular summed locus is not related to the A loci it is not considered further here. The interested reader may refer to the references [6,7]. Alternatively an algorithm for solving nonlinear algebraic equations using the stored A loci information can be employed to solve directly eqs. (6.37) and (6.39) and in addition to check the validity conditions of eqs. (6.38) and (6.40). The only information required in the input data is the relay parameters, the parameters of the various transfer function types listed in Table 6.1, which constitute the partial fraction expansion of G(s), and an initial guess for the solution. Fig. 6.7 shows a graph of the limit cycle frequency, ω, against loop gain, K, obtained from this type of computer program for a relay with $\delta=1$, $h=1$ and $\Delta=0$ and a loop transfer function $G(s) = K/s(1 + s)^2$. The DF solution for the same problem is also shown in the figure. Like the DF solution to the problem two solutions exist for eqs. (6.37) and (6.39) and only the stable one is shown in the figure. The problem of the stability of a solution is discussed in the next section.

In many control problems one wishes to avoid the existence of limit cycles. The above approach can be helpful in these situations since it can be used to determine by how much the system loop gain or phase shift can be changed before a limit cycle will occur. An additional loop phase shift can be obtained by including a time delay in the system transfer function.

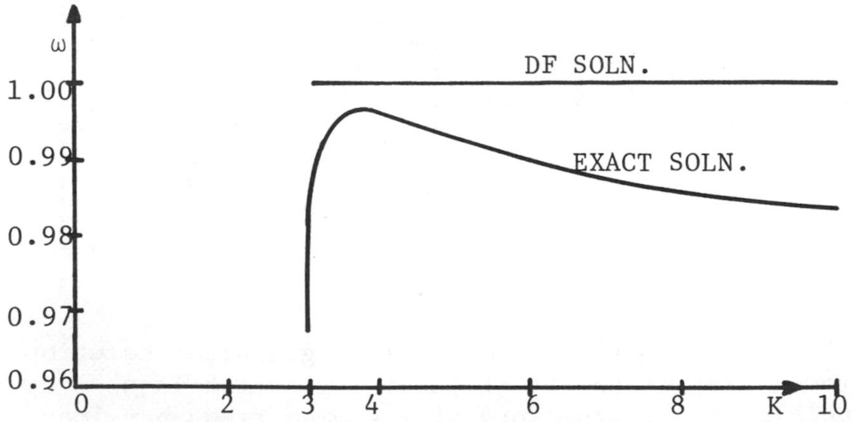

Figure 6.7 Comparison of calculated frequencies

6.5 THE STABILITY OF PREDICTED OSCILLATIONS

The determination of the stability of limit cycle solutions predicted by the procedures of the previous section is not a completely solved problem. When the relay has no dead zone Tsypkin [8, 9] has shown that a necessary condition for stability is that

$$\frac{d\{\text{Im } A_G^\circ(0,\omega)\}}{d\omega} > 0 \qquad (6.43)$$

at the oscillation frequency $\omega=\omega_o$. This result is entirely analogous to the Loeb criterion for the DF method in that it considers a synchronous perturbation. In addition an incremental gain criterion must be satisfied to ensure that an oscillation at an unrelated frequency cannot exist. It can be shown [6] that the exact incremental gain, $N_{i\gamma}$, for a relay with no dead zone is $2h\omega/\pi\dot{x}(0)$. Thus the additional condition that the linear system with an open loop transfer function $N_{i\gamma}G(j\omega)$ should be stable must be satisfied. This will normally be the case unless, for example, the root locus of G(s) has more than two branches in the right hand side of the s plane.

For the relay with dead zone criteria for stability of the predicted oscillations are more complex. When the oscillation is determined using the DF locus and the summed frequency locus $G_m(\theta,\omega)$ application of the Loeb criterion to the solution point appears to give a correct result for synchronous perturbations. Tsypkin [7] has shown that a necessary condition for a solution to be stable is that the angle, ψ, between the two curves giving a solution in Fig. 6.5, should be less than 180°. The angle is measured between tangents drawn at the solution points in the direction of increasing ω and from the tangent to curve (b), the solutions for eq. (6.37), as shown in Fig. 6.5. Nugent [10], on the other hand, has derived a criterion which requires information in addition to the angle, ψ. None of the criteria consider asynchronous perturbations.

The predicted solution is also only valid, as mentioned earlier, if the continuity conditions of eqs. (6.41) and (6.42) are satisfied. These conditions can easily be checked computationally by evaluating the relay input $-c(t)$ from eq. (6.13) with the solution parameters for $\omega\Delta t_1$ and ω substituted. Fortunately this only appears to be necessary when the transfer function G(s) is of a form which will give a highly oscillatory response to the input pulse waveform obtained for the limit cycle solution.

An interesting example [11] of this situation is provided by considering a system with an ideal relay and transfer function $G(s)=(1+2s)/(s+1)(s^2+1)$. The polar locus $G(j\omega)$, which goes to infinity at $\omega=1$ because of the imaginary axis poles, is shown in Fig. 6.8. Since Im $G(j\omega)$ is positive for $\omega<1$ and negative for $\omega>1$ the corresponding A_G° locus, because it involves weighted sums of harmonic frequencies, has Im $A_G^\circ(0,\omega)=0$ for an infinite number of frequencies ω less than unity. The values of these frequencies, using the A loci expressions given in Appendix II for the transfer function types 2, 9 and 10, the latter two with $\zeta=0$, can be shown to be given by

$$\omega = 1/(2n + 0.5) \quad \text{for } n = 1, 2 \ldots\infty . \tag{6.44}$$

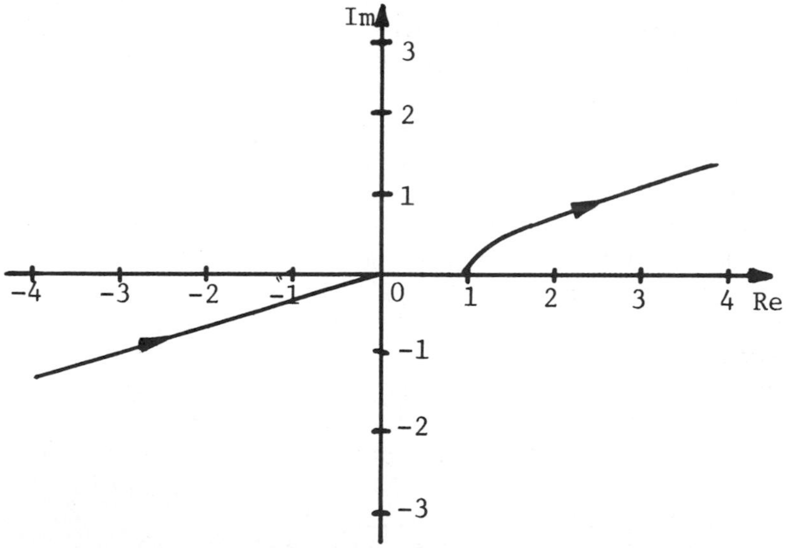

Figure 6.8 Nyquist locus for $G(s) = (1 + 2s)/(s + 1)(s^2 + 1)$

It can further be shown that eq. (6.43) is satisfied for all these solutions. Thus we appear to have a system for which the DF method predicts no limit cycle and the exact method of solution predicts an infinite number of limit cycles. Calculation of the waveform c(t) from eq. (6.13) with ω=0.4, the highest solution frequency, and $\omega\Delta t_1=\pi$ shows that c(t) passes through zero for two values of t in the range $0<t<\pi/0.4$, thus violating the continuity conditions. Simulation of the system shows that it is stable with no limit cycles as predicted directly by the DF method.

The thoughtful reader might question, as a consequence of the oscillatory properties of the above transfer function, whether a symmetrical odd oscillation with the relay switching at intermediate points within the half period might not be possible. Although simulations show this is not the case for the above example such a phenomenon cannot in general be ruled out. Indeed an equivalent situation has been obtained for relays with dead zone and is considered in Section 6.8.

A further aspect regarding applicability of the method, which to the writer's knowledge has not been examined, is its relevance to systems that may have a sliding state. The relay output cannot be correctly represented as a pulse waveform when sliding occurs. Fortunately sliding is only possible for a strictly proper G(s) if the order of the numerator is one less than the denominator so that $\dot{c}(t)$ has discontinuities.

6.6 SYSTEMS WITH MULTIPLE RELAYS

Any limit cycle which exists in an autonomous system with multiple relays and no other nonlinear elements, provided the output waveforms of all the relays may be assumed to be of the same fundamental frequency and of the form shown in Fig. 6.1, can be calculated by the technique presented above. The only additional complication is that the number of nonlinear algebraic equations of the form of eqs. (6.37) and (6.39) which give possible solutions becomes N_1+2N_2, where N_1 is the number of relays without dead zone and N_2 the number of relays with dead zone. If the system structure is such that the output of relay k feeds to the input of relay j via the transfer function $G_{jk}(s)$ the relevant relationships for all relays j=1 to N are

$$\sum_{k=1}^{N} h_k\{A^{\circ}_{G_{jk}}(\alpha_k - \alpha_j,\omega) - A^{\circ}_{G_{jk}}(\gamma_k - \alpha_j,\omega)\} \tag{6.45}$$

has R.P. < 0 and I.P. = $-\pi(\delta_j + \Delta_j)/2$
and

$$\sum_{k=1}^{N} h_k\{A^{\circ}_{G_{jk}}(\gamma_k - \gamma_j,\omega) - A^{\circ}_{G_{jk}}(\alpha_k - \gamma_j,\omega)\} \tag{6.46}$$

has R.P. < 0 and I.P. = $\pi(\delta_j - \Delta_j)/2$
where $\alpha_j = \omega t_j$, $\gamma_j = \omega(t_j+\Delta t_j)$, R.P. denotes real part and I.P. denotes imaginary part. Assuming all the relays have dead zone there are thus 2N equations to solve for the 2N+1 unknowns, ω, α_j and γ_j, for j=1 to N. To obtain a solution one of the angles, say α_1, may be taken equal to zero as the time reference.

This method has been used in reference [5] to determine the exact stability boundary of a relay control system with coulomb friction. Another interesting application is its use to calculate quite accurately limit cycles in feedback systems with saturation type nonlinearities [11]. The saturation nonlinearity is approximated by a stepped waveform, which corresponds to a set of relays with dead zone and no hysteresis in parallel. Obviously the more relays used to approximate the nonlinearity, the larger the number of nonlinear algebraic equations (6.45) and (6.46) to be solved, but the more accurate the resulting solution. The approach can also be used for multivariable systems and further consideration of this topic is deferred to Chapter 7. The determination of the stability of the predicted limit cycle solutions for the general case remains an open question.

6.7 MULTIPLE PULSE OSCILLATIONS

The possibility of obtaining oscillations in relay systems with the relay output waveform different from the basic one shown in Fig. 6.1 has been alluded to in Section 6.5. A more general n-pulse symmetrical odd waveform in shown in Fig. 6.9. The output, c(t),

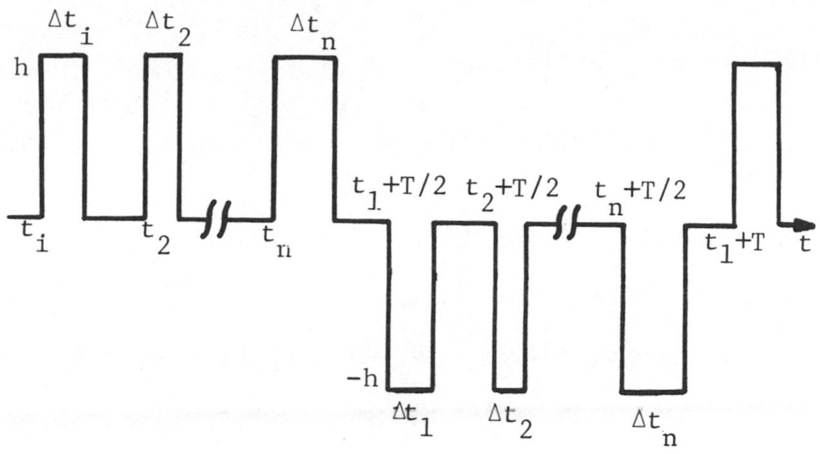

Figure 6.9 n-pulse waveform

of a plant transfer function G(s) with this waveform as input can obviously be written as a summation of eq. (6.13) over all the symmetrical odd pulse train inputs, that is

$$c(t) = (2h/\pi) \sum_{i=1}^{n} \{\mathrm{Im}\, A_G^\circ(-\omega t + \omega t_i, \omega) - \mathrm{Im}\, A_G^\circ(-\omega t + \omega t_i + \omega \Delta t_i, \omega)\}. \qquad (6.47)$$

Similarly, provided $\lim_{s\to\infty} sG(s)=0$, the derivative of c(t) is given by

$$\dot{c}(t) = (2\omega h/\pi) \sum_{i=1}^{n} \{\mathrm{Re}\, A_G^\circ(-\omega t + \omega t_i, \omega) - \mathrm{Re}\, A_G^\circ(-\omega t + \omega t_i + \omega \Delta t_i, \omega)\} . \qquad (6.48)$$

For this oscillation to exist the relay input -c(t) must satisfy the 2n switching conditions

$$-c(t_j) = \delta + \Delta, \; -\dot{c}(t_j) > 0 \qquad (6.49)$$

and

$$-c(t_j + \Delta t_j) = \delta - \Delta, \; -\dot{c}(t_j + \Delta t_j) < 0 \qquad (6.50)$$

for j=1, 2...n. Substituting for c(t) and $\dot{c}$(t) in these conditions leads to the requirement that for this form of oscillation

$$\sum_{i=1}^{n} A_G^\circ(\theta_i - \theta_j, \omega) - A_G^\circ(\theta_i - \theta_j + \Delta\theta_i, \omega) \qquad (6.51)$$

must have R.P. < 0 and I.P. = $-\pi(\delta+\Delta)/2h$
and

$$\sum_{i=1}^{n} A_G^\circ(\theta_i - \theta_j + \Delta\theta_i - \Delta\theta_j, \omega) - A_G^\circ(\theta_i - \theta_j - \Delta\theta_j, \omega) \qquad (6.52)$$

must have R.P. < 0 and I.P. = $\pi(\delta-\Delta)/2h$.
The determination of the parameters of an oscillation with n positive pulses per period thus requires the solution of 2n nonlinear algebraic equations.

This type of oscillation has in fact been found to occur [12, 13] in the system shown in Fig. 6.10, which is a mathematical model of the roll attitude control loop of the Canadian Communications Technology Satellite, CTS. The satellite has large flexible solar panels, which are approximated in the model by the transfer function $G_2(s)=k_1/(s^2+2\zeta s\omega_1+\omega_1^2)$, a very lightly damped second order system in parallel with the rigid body dynamics, and control is obtained by firing jet thrusters. The action of the thrusters, which give very short duration pulses, is modelled by the relay with the pseudo-rate feedback transfer function $G_1(s)$. The transfer function $G_1(s) = K/\{1 + s(\tau_0/\tau_f)\}$ is relay mode dependent, having the time constant τ_0 when the thrusters are on and τ_f when the thrusters are off.

Using eqs. (6.51) and (6.52) oscillations of the form of Fig. 6.9 for several values of n were computed for the system of Fig. 6.11 with a given set of system parameters and $\tau_0=\tau_f$. In addition, the A loci of $G_1(s)$ with $\tau_0\neq\tau_f$ were calculated to obtain results for the system with the mode dependent pseudo -rate controller. The results were in good agreement with simulation studies [13].

This section has been included to emphasize two points. Firstly, that in principle the method of analysis presented can be used to investigate complex

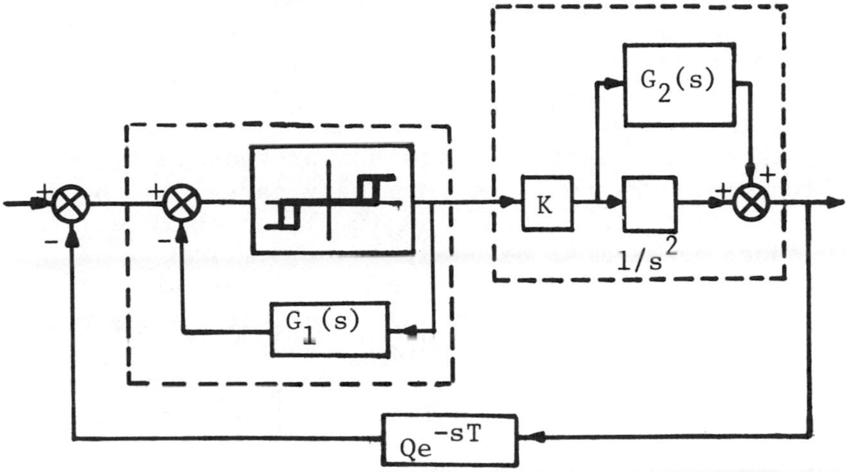

Figure 6.10 Attitude control loop of spacecraft

types of oscillations in relay systems. A difficulty of the method is that one normally must have a reasonable idea of the possible oscillation waveform in order to provide suitable initial estimates for the required solutions to the nonlinear algebraic eqs. (6.51) and (6.52). Secondly, experience suggests that this type of oscillation will not occur regularly, but its possibility cannot be ruled out in systems where G(s) has lightly damped poles.

6.8 ASYMMETRICAL OSCILLATIONS

Asymmetrical oscillations occur in many feedback systems most frequently where the nonlinearity is not odd symmetric or the system is subject to bias inputs. The set point in the on-off temperature control system, for example, simply changes the average heat output by varying the relay on-off ratio. The input waveform to the plant transfer function, when a limit cycle exists, can again be expressed in terms of a summation of the waveforms $y_i(t)$ given by eq. (6.1). To generate the assumed waveform the relay input x(t) will again have to satisfy certain switching conditions. For the simplest case, namely an on-off relay without dead zone, and a one pulse mode, $y_1(t)$, two switching conditions will exist since in general $\omega\Delta t_1 \neq \pi$. Thus two nonlinear algebraic equations have to be solved to obtain the limit cycle parameters. This doubling of the number of equations which yield the solution applies to the investigation of asymmetrical oscillations in all the situations considered earlier. Also all terms, not just odd, are now present in the series which yield the real and imaginary parts for the A_G loci given by eqs. (6.6) and (6.7), so that Table II.2 in Appendix II must be used.

Because of the increased number of unknowns a graphical solution is now only possible if the relay has no dead zone and the relay output has one pulse per period, that is n=1. Consider the system of Fig. 1.1 to have a constant input r(t)=R, and a relay with $\delta=0$ and output levels 0 and h. The input to the relay, x(t), is given by

$$x(t) = R - c_1(t) \tag{6.53}$$

where $c_1(t)$ is given by eq. (6.8) with i=1 and the switching conditions, assuming the relay switches positive at time $t_1=0$, are

$$x(0) = \Delta \ , \ \dot{x}(0) > 0 \tag{6.54}$$

and

$$x(\Delta t_1) = -\Delta \ , \ \dot{x}(0) < 0 \tag{6.55}$$

Substituting the values of x(t) in these equations gives for the limit cycle solution

$$A_G(0,\omega) - A_G(2\pi\rho,\omega) \tag{6.56}$$

must have R.P. < 0 and I.P. = $(\pi/h)\{(R-\Delta)-\rho h G(0)\}$ and

$$A_G(0,\omega) - A_G(-2\pi\rho,\omega) \tag{6.57}$$

must have R.P. < 0 and I.P.=$-(\pi/h)\{(R+\Delta)-\rho h G(0)\}$ provided $\lim_{s\to\infty} sG(s)=0$ otherwise corrections have to be made for the discontinuity in $\dot{x}(t)$ at the switching instants as discussed earlier. Here $\rho=\Delta t_1/T=\omega\Delta t_1/2\pi$.

As a specific example [14] a temperature control system having $G(s)=Ke^{-s\tau}/(1+s\tau_1)(1+s\tau_2)$ with $\tau_1=1$ s., $\tau_2=4$ s., $\tau=0.0833$ s. and $K=0.025°F/W$. and a heater characterized by $\Delta=0$ and $h=1000$ W. is considered. For a reference input of $R=68°F$ the loci of eqs. (6.56) and (6.57) are shown plotted for various values of ρ in Fig. 6.11. The set of possible solutions for the two loci are shown in the $\rho-\omega$ plane of Fig. 6.12 from which it is seen that the limit cycle solution has $\rho=0.278$ and $\omega=3.25$ rad/s. This corresponds to a regulated output average temperature of 70°F.

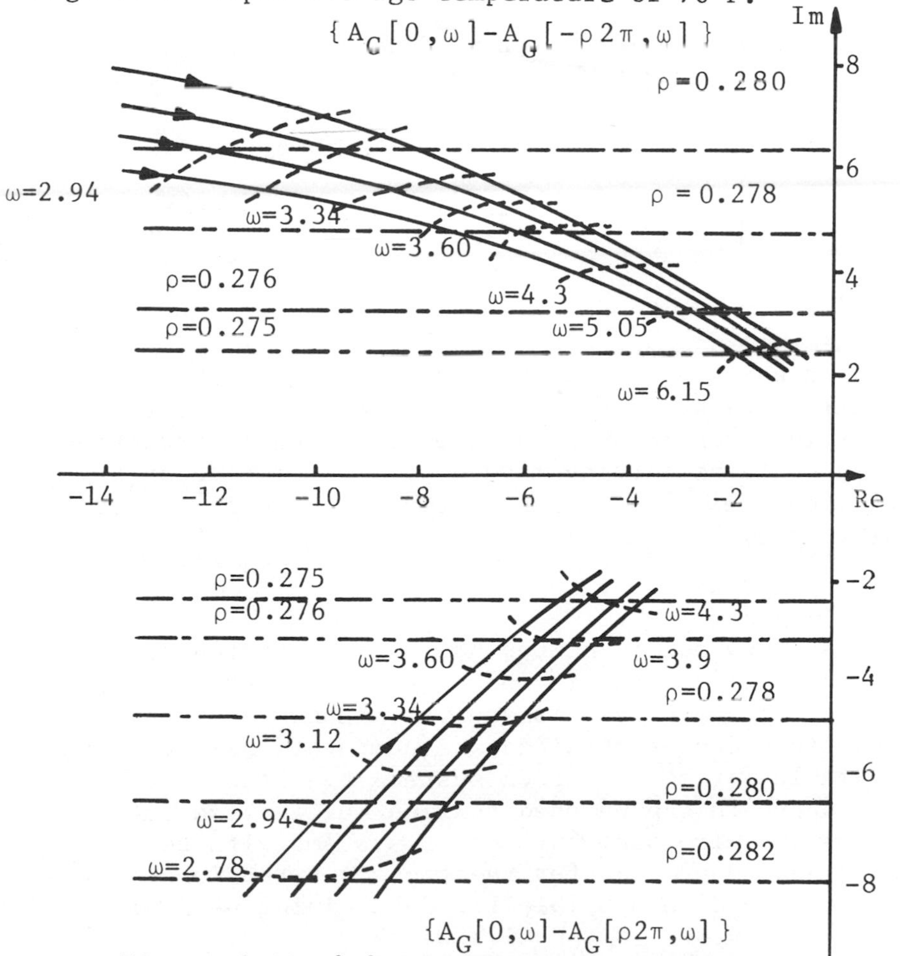

Figure 6.11 'A' loci for temperature control system

Computational solutions can be obtained when more than two nonlinear algebraic equations have to be

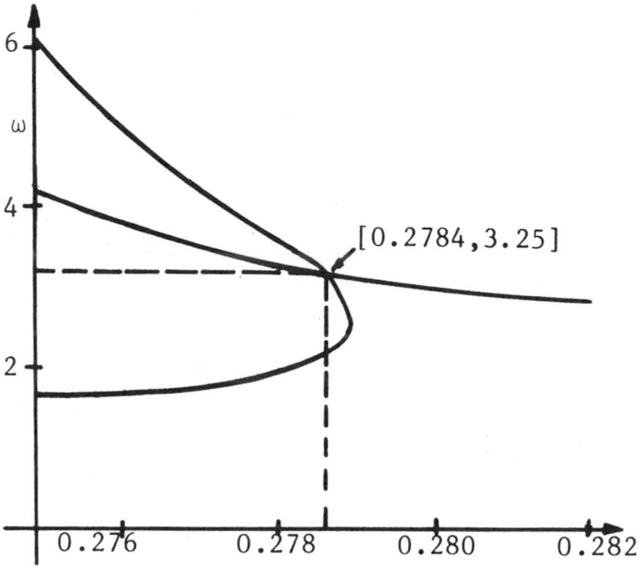

Figure 6.12 Solution curves from Fig. 6.11

solved using a general program similar to that for determining symmetrical oscillations. In addition solutions can be obtained when the transfer function G(s) contains a relay mode dependent time delay or time constant [14] which is typical of many thermal processes.

6.9 SUMMARY

A frequency domain method has been presented for investigating periodic modes in relay systems. For some elementary systems analytical and/or graphical solutions can be obtained. A computational technique suitable for more complex systems has been outlined. The approach can be used to find multi-pulse oscillations and also oscillations in systems with relay mode dependent transfer functions. Extensions of the method are also possible for determining periodic modes in systems with pulse width modulation and/or pulse frequency modulation [14]. New results on oscillation stability have recently been obtained [15].

REFERENCES

1. Hamel, B.: "Contribution a l'etude mathematique des systemes de reglage par tout-ou-rien", C.E.M. V.: Service technique Aeronautique, 1949, 17.

2. Bohn, E.V.: "Stability margins and steady state oscillations in on-off feedback systems", I.R.E. Trans. Circuit Theory, 1961, CT-8, pp. 127-130.

3. Chung, J.K-C. and Atherton, D.P.: "The determination of periodic modes in relay systems using the state space approach", Int. J. of Control, 1966, 4, pp. 105-126.

4. Tsypkin, J.S.: "Theorie der relais systeme der automatischen regelung", R. Oldenbourg-Verlag, Munich, 1958.

5. Atherton, D.P.: "Conditions for periodicty in control systems containing several relays", Proc. I.F.A.C. Congress, Paper 28E, London, 1966.

6. Turnbull, G.F., Atherton, D.P. and Townsend, J.M.: "A method for the theoretical analysis of relay amplifiers", Proc. I.E.E., 1965, 112, pp. 1039-1055.

7. Atherton, D.P.: "Nonlinear control engineering", Van Nostrand Reinhold, London, 1975, Chapter 6.

8. Gille, J.C., Pelegrin, M.M. and Decaulne, P.: Feedback control systems, McGraw-Hill, New York, 1959, Chapter 26.

9. Tsypkin, J.S.: "Relay automatic control systems", Moscow, 1974.

10. Nugent, S.T.: "Steady state oscillations in multivariable relay control systems", Ph.D. thesis, University of New Brunswick, Canada, 1967.

11. Atherton, D.P. and Ramani, N.: "A Method for the evaluation of limit cycles", Int. J. of Control, 1975, 21, pp. 375-384.

12. Millar, R.A. and Vigneron, F.R.: "Attitude stability of flexible spacecraft which use dual time constant feedback lag network pseudo-rate control", AIAA/CASI 6th Communications Satellite Systems Conference, Montreal, p. 266, 1976.

13. Rao, U.M. and Atherton, D.P.: "Multi-pulse oscillations in relay systems", Proc. 7th IFAC World Congress, Helsinki, Vol. 3, Paper No. 42.4, pp. 1747-1754, Pergamon Press, 1978.

14. Rao, U.M.: "Studies in satellite attitude control system design", Ph.D. thesis, University of New Brunswick, Canada, 1979.

15. Balasubramanian, R.: "Stability of limit cycles in feedback systems containing a relay" to be published in IEE Proceedings-D.

PROBLEMS

1. Compare the limit cycle amplitude and frequency solutions obtained by the DF and exact methods for a system with an ideal relay with h = 1 and $G(s) = (1 - s)/(1 + s)^2$.

2. Repeat problem 6.1 for $G(s) = e^{-2s}/(1 + s)$ and compare the two solutions.

3. Obtain exact solutions for the example considered in section 4.9.

4. A feedback system has a controller, consisting of a relay with $\Delta=0$, $\delta=1$ and h=10 in parallel with a linear gain of unity, and a plant transfer function of $K/s(s+1)^2$. Find the maximum value of K for stability by the DF and exact methods.

5. Show analytically that the Tsypkin locus for $G(s) = (as + b)/(s + 2)(s^2 + 1)$ has multiple crossings of the real axis. For what values of a and b can the Tsypkin conditions for a stable limit cycle with an ideal relay controller be satisfied? Does the limit cycle exist?

6. Evaluate by the DF and exact methods any limit cycle in a system with an ideal relay with h=1, and $G(s) = K/(s + a)^2(s - 1)$ for (i) a=4 and (ii) a=3. Is there any difference in the form of the limit cycle for the two cases.

7. A position control system may be assumed to have an ideal relay controller with h=1 and a plant transfer function of $K/s(s + 2)$. Coulomb friction in the system can be represented by a feedback loop around the plant consisting of a differentiator and ideal relay with h=0.5. Evaluate any limit cycle in the system.

8. A temperature control system has a relay controller with, $\delta=0$, $\Delta=1$, and h=10. The process can be represented by the transfer function $8e^{-2s}/(s + 1)$. Evaluate the limit cycle

frequency, output amplitude and mean value of the output for reference input levels of 40 and 50 respectively. (Relay on off levels are h and 0)

9. Evaluate the exact incremental gain N_i for (a) an ideal relay and (b) a relay with dead zone when its unperturbed output is a symmetrical oscillation with one pulse per period. Show how the result can be written in terms of the A locus of G(s) when the relay output corresponds to a limit cycle in the feedback loop of Fig. 1.1.

10. Repeat question 9 parts (a) and (b) assuming the relay has hysteresis.

11. Solve for the limit cycle in Example 4 of chapter 5 and compare the solution with that obtained using the SBDF.

CHAPTER 7
Multivariable Systems

7.1 INTRODUCTION

It is possible to extend many of the methods considered in earlier chapters for investigating the stability of the nonlinear system of Fig. 1.1 to the multivariable system shown in Fig. 7.1. Although some of the techniques we will discuss may be used for other configurations our prime concern will be directed to results for the block diagram of Fig. 7.1. The most general form of the nonlinearity N, which we will assume to have an equal number of inputs and outputs, is

$$y_i = n_i(x) \tag{7.1}$$

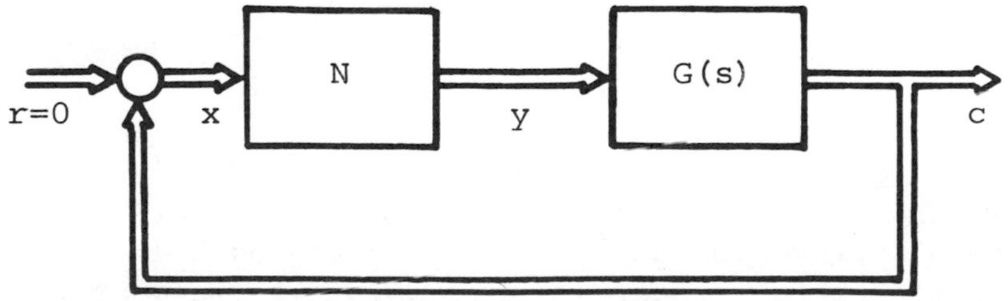

Figure 7.1 Nonlinear multivariable system

where the input and output vectors x and y have components x_i and y_i, $i=1,2,\ldots,n$, respectively. Two particular forms of the nonlinearity N often occur in practical situations. These are where N consists of individual nonlinear characteristics so that eq. (7.1) may be written

$$y_i = \sum_{j=1}^{n} n_{ij}(x_j) \tag{7.2}$$

and also when N is diagonal, so that in addition $n_{ij}(x_j)=0$ for $j \neq i$, and

$$y_i = n_{ii}(x_i) = n_i(x_i) \tag{7.3}$$

For this case N will be referred to as a diagonal nonlinearity.

Before considering the nonlinear problem it is appropriate to review briefly techniques for determining the stability of the linear system obtained when N is replaced by the linear gain matrix K. In this case

$$C(s) = G(s)K\{R(s) - C(s)\}$$

giving

$$C(s) = F^{-1}(s)Q(s)R(s) \tag{7.4}$$

with

$$F(s) = I + Q(s) \tag{7.5}$$

and

$$Q(s) = G(s)K \tag{7.6}$$

Q(s) and F(s) are respectively the return ratio and return difference matrices for the feedback loop measured at the linear gain input. Eq. (7.4) may also be written

$$C(s) = \frac{\text{adj}\{F(s)\}Q(s)R(s)}{\det F(s)} \tag{7.7}$$

which suggests that stability is related to the properties of det F(s). It can be shown that [1]

$$\frac{\Phi_c(s)}{\Phi_o(s)} = \frac{\det F(s)}{\det F(\infty)} \tag{7.8}$$

where $\Phi_o(s)$ and $\Phi_c(s)$ are the open and closed loop characteristic polynomials, respectively. The closed loop system is stable when all the zeros of $\Phi_c(s)$ lie in the left hand side of the s plane. Thus, if the open loop system is stable, the stability of the closed loop system can be found from the roots of the characteristic equation

$$\det F(s) = 0 \tag{7.9}$$

For a specific case stability can therefore be investigated, for example, by using the Hurwitz Routh criterion on eq. (7.9) or if eq. (7.9) is written in the form $1 + H(s) = 0$, by plotting the polar frequency response locus of H(s).

When the open loop transfer matrix is obtained from a state space description, that is

$$\dot{x} = Ax + Bu \tag{7.10}$$

$$y = Cx + Du$$

with

$$Q(s) = C(sI - A)^{-1}B + D \tag{7.11}$$

then $\Phi_o(s) = \det(sI - A)$ and the roots of $\Phi_o(s)$ can be easily determined. If Q(s) is given directly as a transfer function matrix then it can be shown that

$$\det(sI - A) = p(s) \tag{7.12}$$

when eqs. (7.10) correspond to a controllable and observable system. p(s), the pole polynomial of Q(s), is the least common denominator of all the non-zero minors of all orders of Q(s).

Several variations of graphical stability criteria can be proved from eq. (7.8) for the general case where $\Phi_o(s)$ has P right half plane zeros [2]. They are summarised in the following theorems

Theorem 1

The system S is asymptotically stable if the Nyquist plot of det $\{I + G(s)K\}$ [or det $\{I + KG(s)\}$] encircles the origin P times counterclockwise.

Since

$$\det\{I + KG(s)\} = \prod_{i=1}^{m} 1 + \lambda_i(s) \qquad (7.13)$$

where $\lambda_i(s)$ are the eigenvalues of KG(s) Theorem 2 follows.

Theorem 2

The system S is asymptotically stable if the sum of the counterclockwise encirclements of (-1,j0) by the Nyquist plots $\lambda_i(s)$ is P.

The Nyquist plots $\lambda_i(s)$, $s=j\omega$, are known as characteristic loci. In particular it should be noted that if $K = kI$ then the stability of the system as k varies can be determined from the characteristic loci of G(s). Theorem 1 can also be expressed in terms of the inverse Nyquist locus of G(s), denoted by $\hat{G}(s)$, where $\hat{G}(s) = \text{adj}\, G(s)/\det G(s)$.

Theorem 3

The system S is asymptotically stable if the Nyquist plot of det $\{\hat{G}(s) + K\}$ encircles the origin a net total of P times counterclockwise more that the plot of det $\hat{G}(s)$.

Theorem 4

The system S is asymptotically stable if the Nyquist plot of det $\{G(s) + \hat{K}\}$ encircles the origin P times counterclockwise.

The above theorems apart from Theorem 4 are also true if $K = K(s)$.

Although it is not difficult to make use of the above theorems to assess the stability of a particular system it is not always easy to use them to investigate how the stability of the system may change with variations in system parameters. Several theorems based on the diagonal dominance properties of the transfer matrix $G(s)$ are known which can be more helpful in some instances for this situation.

A complex nxn matrix Z, with elements z_{ij}, is said to be diagonal dominant if its elements satisfy one or other of the following sets of conditions, where the primed summation indicates that the term $j=i$ is to be omitted:

Row dominance

$$|z_{ii}| > \sum_{j=1}^{n}{}' |z_{ij}| \quad (i = 1,2,\ldots n) \tag{7.14}$$

Column dominance

$$|z_{ii}| > \sum_{j=1}^{n}{}' |z_{ji}| \quad (i = 1, 2,\ldots n) \tag{7.15}$$

Mean dominance

$$|z_{ii}| > \sum_{j=1}^{n}{}' \{|z_{ij}| + |z_{ji}|\}/2 \quad (i = 1,2,\ldots n) \tag{7.16}$$

The significance of these conditions for stability theory is found in the following Lemma.

Lemma 1

If a rational function matrix $Z(s)$ is diagonal dominant for every s on D, the Nyquist contour in the s plane, then the number of encirclements of the origin by the Nyquist plot of det $Z(s)$ is equal to the sum of the number of encirclements by the Nyquist plots of the diagonal elements, $Z_{ii}(s)$, of $Z(s)$. This leads to the following two theorems:

Theorem 5

If the Nyquist plot of $G_{ii}(s)$ encircles the point $(-k_i^{-1}, j0)$, n_i times counterclockwise, and if for each s on D

$$|G_{ii}(s) + k_i^{-1}| > \Delta_i(s), \quad i = 1, 2 \ldots n \tag{7.17}$$

then the closed loop system S is asymptotically stable, if and only if,

$$\sum_{i=1}^{n} n_i = P \tag{7.18}$$

Here $G_{ij}(s)$ are the elements of $G(s)$, $K = \text{diag}(k_i)$ and $\Delta_i(s)$ is defined from the off diagonal elements of $G(s)$ by

$$\Delta_i(s) = \sum_{j=1}^{n}{}' |G_{ij}(s)| \tag{7.19}$$

or

$$\Delta_i(s) = \sum_{j=1}^{n}{}' |G_{ji}(s)| \tag{7.20}$$

or

$$\Delta_i(s) = \sum_{j=1}^{n}{}' \{|G_{ij}(s)| + |G_{ji}(s)|\}/2 \tag{7.21}$$

Theorem 6

If the Nyquist plot of $\hat{G}_{ii}(s)$ encircles the point $(-k_i, j0)$, $\hat{n}_i$ times counterclockwise, and the origin N_i times counterclockwise, and if, for each s on D,

$$|\hat{G}_{ii}(s) + k_i| > \hat{\Delta}_i(s), \quad i = 1, 2 \ldots n \tag{7.22}$$

and

$$|\hat{G}_{ii}(s)| > \hat{\Delta}_i(s), \quad i = 1, 2 \ldots n \tag{7.23}$$

where $\hat{\Delta}_i(s)$ is defined as in eqs. (7.19)–(7.21) with respect to the off diagonal elements of $\hat{G}(s)$, then the system S is asymptotically stable, if and only if,

$$\sum_{i=1}^{n} (\hat{n}_i - N_i) = P \qquad (7.24)$$

Here $\hat{G}_{ij}(s)$ are the elements of $\hat{G}(s)$ which are not normally equal to $\{G_{ij}(s)\}^{-1}$.

7.2 ABSOLUTE STABILITY CRITERIA

Although a few results on absolute stability have been obtained for the nonlinearity N non-diagonal [3,4,5] their applicability due to the conservativeness of the results appears very limited. We therefore restrict our considerations here to N given by eq. (7.3). As in chapter 3 sector and slope bounds on the nonlinearities will be used, thus if $n_i(x) \in [k_{i1}, k_{i2}]$ then $k_{i1} \leq n_i(x_i)/x_i \leq k_{i2}$. Since several of the known graphical stability criteria can be proved from the Jury and Lee criterion [6], which is analogous to the Popov criterion for the single variable system, we begin with this result.

7.2.1 Jury and Lee criterion

Theorem 7

The system of Figure 7.1 is asymptotically stable if there exists a real diagonal matrix Q such that the matrix P where

$$P = A + A^* \qquad (7.25)$$

$$A = (I + j\omega Q)G(j\omega) + \text{diag}\{k_i^{-1}\} \qquad (7.26)$$

is positive definite. The nonlinearities are assumed to be single valued with $n_i(x_i) \in [0, k_i]$ and * denotes the conjugate transpose.

A difficulty with this method is that if Q = diag (q_i) we have n free parameters q_i which can be varied in order to try and obtain a positive definite P. Two possibilities, which give weaker results, namely choosing Q = qI or Q = 0 are obvious simplifications and are considered in the next sections. It is possible when the k_i are given to compute a suitable Q if one exists or when the k_i are unknown to determine a Q which maximizes, say

$$\sum_{i=1}^{n} k_i .$$

This approach, however, gives no physical understanding and is of little use for design purposes.

Using Gerschgorin's theorem on the bounds for eigenvalues conditions for P to be positive definite can be found and the results may be expressed in the following two theorems [7].

Theorem 8

The system of Fig. 7.1 is asymptotically stable if for all i, i=1,2...n, the area defined by

$$\tilde{G}_{ii}(j\omega) + d_i(\omega) \exp (j\phi), \ \phi = 0 \text{ to } 2\pi, \ \forall\omega$$

lies to the right of the straight line of slope q_i^{-1} drawn through the point $(-k_i^{-1}, j0)$, where $\tilde{G}_{ii}(j\omega)$ is the Popov locus of $G_{ii}(j\omega)$ and

$$d_i(\omega) = (1/2)\{|1+j\omega q_i|[\sum_{j=1}^{n}{}' |G_{ij}(j\omega)|] +$$

$$\sum_{j=1}^{n}{}' \{|1+j\omega q_j||G_{ji}(j\omega)|\} \tag{7.27}$$

Theorem 9

The system of Fig. 7.1 is asymptotically stable for all i, i=1,2...n if

$$\text{Re}\{(1+j\omega q_i)G_{ii}(j\omega)\} + \{k_i^{-1} - \ell_i(\omega)\} > 0, \ \forall\omega$$

where $$\ell_i(\omega) = (1/2) \sum_{j=1}^{n}{}' |p_{ij}| \tag{7.28}$$

$$p_{ij} = (1+j\omega q_i)G_{ij}(j\omega) + (1-j\omega q_j)G^*_{ji}(j\omega) \tag{7.29}$$

Both the above theorems have graphical interpretations. The system is stable by Theorem 8 if the Popov loci of the diagonal elements banded with circles of radius $d_i(\omega)$ lie to the right of the respective Popov lines and by Theorem 9 if the loci $M_i(j\omega)$ obtained by displacing $G_{ii}(j\omega)$ by $\ell_i(\omega)$ in the negative direction lie to the right of the respective Popov lines. Since the $M_i(j\omega)$ lie within the banded loci Theorem 9 gives the best results. A difficulty in using the above results even with an interactive graphics terminal is that $d_i(\omega)$ and $\ell_i(\omega)$ depend upon all the q_i, i=1, 2...n. The choice of the best q_i, say to maximize

$\prod_{i=1}^{n} k_i$, is thus one of trial and error, although for n=2 or 3 examination of the relative magnitudes of the off diagonal elements of G(jω) gives a good indication of appropriate values of q_i.

Table 7.1 gives results obtained by the above methods for the allowable sector widths for stability of a system with

$$G(s) = \begin{pmatrix} \dfrac{1.0}{(s+1)(s+2)(s+3)} & \dfrac{0.3}{(s+2)(s+3)} \\ \dfrac{1.5}{(s+3)(s+4)} & \dfrac{1.0}{(s+4)(s+5)} \end{pmatrix}$$

TABLE 7.1

Stability Results

		Theorem 8		Theorem 9	
q_{11}	q_{22}	k_1	k_2	k_1	k_2
0	0	9.68	26.67	12.2	26.67
0.2	0.2	10.7	25.0	11.75	26.67
0.5	0.5	9.08	20.0	10.82	26.67
0.5	0	11.82	24.21	17.24	26.67
1.0	0	-	-	18.95	26.67
1.1	0	-	-	18.58	26.67

Fig. 7.2 shows a plot of the relevant loci for loop 1 with $q_i = 0.5$ and $q_2 = 0$.

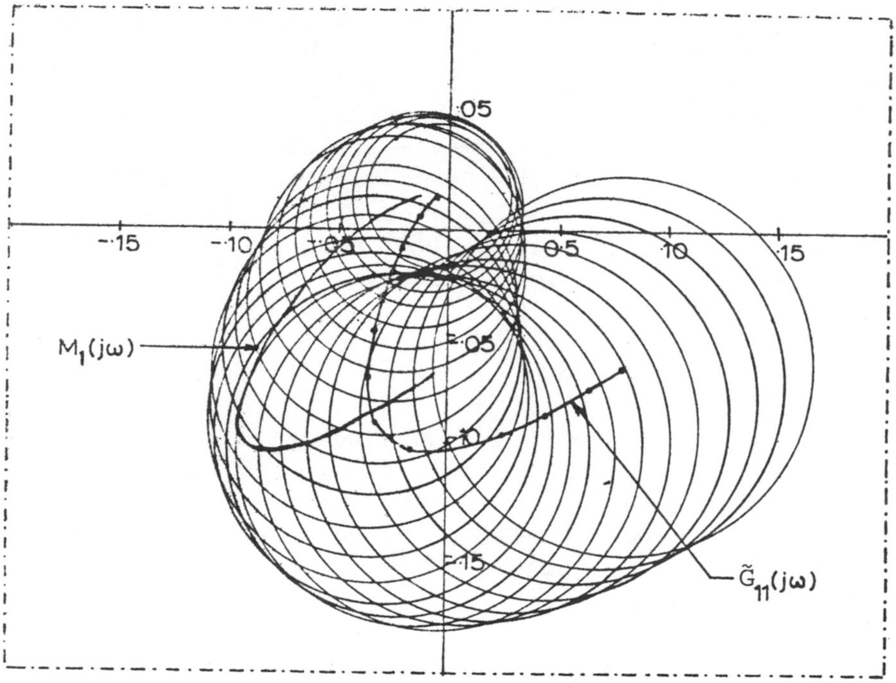

Figure 7.2 Loci for stability investigation

Two other graphical stability criteria based on choosing Q=qI are given in reference [8]. Both the criteria use results from matrix algebra on the field of values of a complex matrix, which it is known contains the eigenvalues of the matrix [9]. For a given choice of q,q=0 being particularly convenient, Popov type criteria are applied to banded loci of either the diagonal elements or eigenvalues of G(jω). Unlike the Gerschgorin procedures the bands are obtained by forming rectangles not circles around the loci.

7.2.2 Circle criteria

A circle criterion can be obtained from the results of the previous section taking Q=0 and using a multivariable pole transformation analogous to that discussed in Section 3.3.1 for the single variable system. The resulting criterion, which was proved independently by Cook [10], requires banded Nyquist loci of the diagonal elements $G_{ii}(j\omega)$ of $G(j\omega)$ to avoid the discs $D(k_{i1},k_{i2})$ for all $i=1,2...n$. The radii, $d_i'(\omega)$, of the circular bands on $G_{ii}(j\omega)$ are given by eq. (7.27) with $Q=0$, that is

$$d_i'(\omega) = (1/2) \sum_{j=1}^{n}{}' \{|G_{ij}(j\omega)| + |G_{ji}(j\omega)|\} \quad (7.30)$$

A similar result has been obtained by Rosenbrock [11] which uses circular bands proportional to row or column elements rather than the mean dominance condition of eq. (7.30). The critical discs, however, have to be enlarged by certain factors in this criterion so that the results are usually more conservative.

Cook [12] has also proved a multivariable version of the off axis circle criterion for $n_1'(x)\in[m_1,m_2]$, but the condition only ensures the absence of limit cycles not absolute stability. The Nyquist loci $G_{ii}(j\omega)$ again banded according to eq. (7.30) are required to avoid the off axis discs $D'(m_1,m_2)$ for all i. All the above circle criteria can be expressed in terms of the elements of $\hat{G}(j\omega)$ instead of $G(j\omega)$, in which case the critical discs will pass through $(-k_{i1},j0)$, $(-k_{i2},j0)$ rather than $(-k_{i1}^{-1},j0)$, $(-k_{i2}^{-1},j0)$ and the requirement for stability is that the inverse Nyquist loci encircle the disc the same number of times they encircle the origin.

A stability criterion with a simple graphical interpretation is given in the following theorem due to Falb et al [13].

Theorem 10

If the system *S* is such that $G(s)$ is normal that is $G(s)G^*(s) = G^*(s)G(s)$ and open loop stable, and N is a diagonal nonlinearity with all $n_i(x_i)\in(k_1,k_2)$ then the system is asymptotically stable if the eigenvalue loci of $G(j\omega)$ do not enclose the critical disc $D(k_1,k_2)$.

The practical applicability of this result is strictly limited by the constraint that $G(s)$ must be normal. Mees and Rapp [14], however, have shown that banded eigenvalue loci can be used when G(s) is not normal; the width of the bands around the loci being dependent on a measure of the departure of G(s) from normality.

7.2.3 Limit cycle criteria

A result similar to that given in section 3.7 for the single variable system has recently been obtained for the multivariable case [15]. It is shown that provided for large inputs the nonlinearities approach gains within the sectors for which the linear system is stable then the response of the nonlinear system of Fig. 7.1 cannot go unbounded. This result indicates, for example, that the off axis criterion [12] can normally be used as a stability criterion.

When absolute stability cannot be guaranteed several results similar to those for the single variable case have been obtained which rule out the possibility of limit cycles with particular frequencies [16,17]. Basically the results show that Theorems 12 and 13 of chapter 3 can be used with respect to the banded frequency loci of Theorem 8 or the shifted frequency loci, $M_i(j\omega)$, of Theorem 9. The results can also be extended to include the slope as well as the sector bounds of the nonlinearities but their application is complicated by the addition of further free parameters.

7.3 LINEARIZATION

Although we have seen that Aizerman's conjecture is not in general true for a single variable system, the types of transfer functions which violate the conjecture are not often representative of control system dynamics. It is therefore appropriate to briefly examine linearization for the multivariable case. N is taken to be an uncoupled diagonal nonlinearity and to simplify the initial discussion a two variable system with $n_i(x_i) \in [0,k]$ for i=1,2 will be assumed. When $n_i(x_i)$ is replaced by the linear gain K_i, i=1,2 it is possible to compute the stability boundary of the linear system in the $K_1 - K_2$ plane, as illustrated in Fig. 7.3 for positive values of K_1 and K_2.

The equivalent of the Aizerman conjecture for the multivariable system of Fig. 7.1 can be stated as

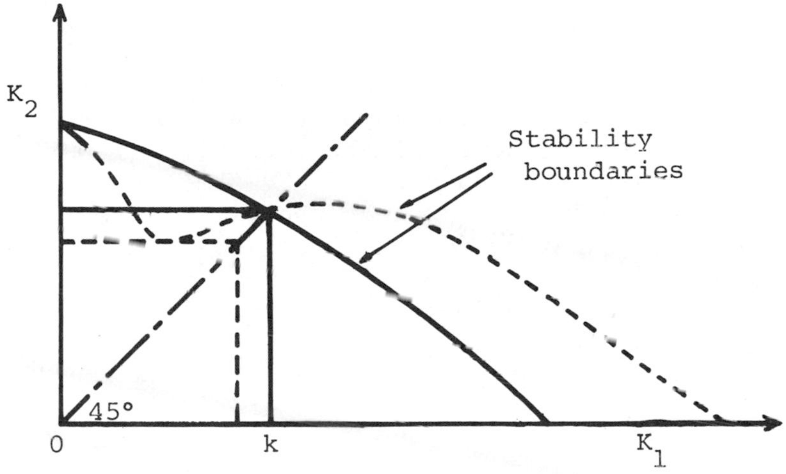

Figure 7.3 Gain plane stability bounding for linear system

follows:

The system of Fig. 7.1 will be asymptotically stable for all $n_i(x_i) \in [0,k]$ if the equivalent linear system is stable for all $K_i \in (0,k)$.

For the two variable system the maximum value of k can be found from the largest square which fits inside the stability boundary as shown in Fig. 7.3. Since a significant computational time may be required to determine the shape of the stability boundary, or surface in higher dimensions, it would be valuable to have conditions which G(s) must satisfy so that the largest square can easily be found; either from the allowable uncoupled loop gains or using eigenvalue methods, as is illustrated for the full, but not for the dotted curves, in Fig. 7.3. Unfortunately no results have been obtained on this problem, apart from those which can be deduced from the stability results of the previous section. For example, as a consequence of theorem 10, it is obvious that one can find k from the eigenvalues of G(s) when G(s) is normal.

For all practical purposes as with the single variable system, the result of reference [15] given in the previous section means that if the words 'asymptotically stable' in the multivariable version of the Aizerman conjecture are replaced by 'asymptotically stable or possess a finite amplitude oscillation' the conjecture is valid.

7.4 DF GRAPHICAL METHODS

When the signals x_i at the input to the nonlinearity N, which is assumed to be of the form given by eq. (7.2), are taken as sinusoids of the same frequency with complex amplitudes a_i then the first harmonic balance or DF equation for the loop is

$$\{I + G(j\omega)N(a)\}a = 0 \qquad (7.31)$$

if the loop is considered opened at the nonlinearity input, or

$$\{I + N(a)G(j\omega)\}b = 0 \qquad (7.32)$$

if the loop is considered opened at the nonlinearity output, where b_i is the first Fourier coefficient of the nonlinearity output y_i.

A sufficient condition for these equations to have a non-zero solution is that

$$\det\{G(j\omega) + \hat{N}(a)\} \neq 0 \qquad (7.33)$$

or

$$\det\{\hat{G}(j\omega) + N(a)\} \neq 0 \qquad (7.34)$$

A few simple graphical DF methods [18,19,20] are known which can be used to check expressions (7.33) and (7.34). The results are often conservative, however, since if the graphical criteria are violated all one can conclude is that a limit cycle may exist. This situation occurs because if a limit cycle solution exists it is for specific values of ω and a, whereas the expressions (7.33) and (7.34) are checked over the allowable ranges of a and ω, rather than for a consistent set.

Mees [18,21] used Gerschgorin's theorem on eigenvalue bounds to show that when N is a diagonal nonlinearity Theorems 5 and 6 can be used with k_i replaced by $N_i(a_i)$. The graphical interpretation of these theorems is that the $G_{ii}(j\omega)$ loci banded with circles of radii $\Delta_i(j\omega)$ {or $\hat{G}_{ii}(j\omega)$ loci banded with circles of radii $\hat{\Delta}_i(j\omega)$} should have the correct number of encirclements of $-1/N_i(a_i)$ {or $N_i(a_i)$ and the origin} for i=1,2,...,n. When N satisfies eq. (7.2) a simple graphical interpretation is only possible using the inverse Nyquist diagram approach of expression (7.34) and $\hat{\Delta}_i(j\omega)$ computed over the columns of $\hat{G}(j\omega)$ according to eq. (7.20). In this case the banded $\hat{G}_{ii}(j\omega)$ loci must have the correct number of encirclements of $-N_{ii}(a_i)$, banded with circles of radius

$$\sum_{j=1}^{n}{}' \; |N_{ji}(a_i)| \quad \text{for all i, and the origin.}$$

A graphical interpretation of a result due to Ramani [20,22,23] using a theorem of Hirsch [24] requires only one plot for an n variable system when N is a diagonal nonlinearity to check either condition (7.33) or (7.34). For condition (7.33) a banded polar frequency response locus formed by rectangles at each frequency with corner points $\lambda_1+j\lambda_3$, $\lambda_1+j\lambda_4$, $\lambda_2+j\lambda_3$ and $\lambda_2+j\lambda_4$ must not encircle a rectangle containing the DFs of the nonlinearities. λ_1 and λ_2 are the minimum and maximum eigenvalues of $\{G(j\omega) + G^*(j\omega)\}/2$ and λ_3 and λ_4 are the minimum and maximum eigenvalues of $\{G(j\omega) - G^*(j\omega)\}/2j$.

To illustrate the procedure outlined above we consider the stability of the system of Fig. 7.1 with N a diagonal nonlinearity consisting of two identical relays with dead zone for which h=1, δ=1 and Δ=0 and

$$G(s) = \frac{K}{s(s+1)^2}\begin{pmatrix} 1 & 0.3 \\ -0.2s-0.2 & 1 \end{pmatrix}$$

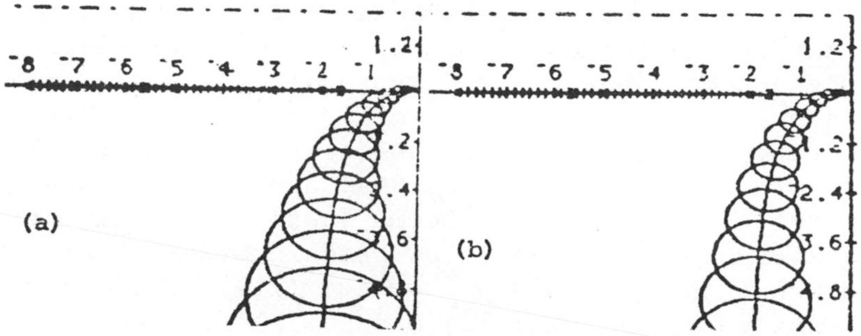

Figure 7.4 Mees plot for example (a) i=1 (b) i=2

Fig. 7.4 shows the Mees plots for the two loops using row dominance with K=1 and Fig. 7.5 the Hirsch plot for the same system. The maximum values of K for stability from these plots are approximately 1.50 and 1.25, respectively. Fig. 7.6 shows the stability boundary for G(s) in the K_1-K_2 gain space discussed in the previous section. The maximum square which fits inside the stable region of the gain space, which in this case can be calculated from the characteristic loci, has sides of 1.14. Thus since the relays lie in the sector $[0,2/\pi]$ the Aizerman conjecture gives a maximum gain K of 1.79 for stability. The exact solution for the maximum value of K for stability given by the Tsypkin method discussed in Section 7.6 is 2.25.

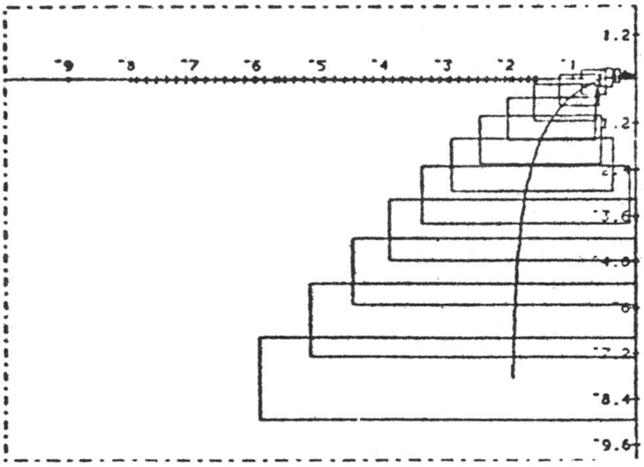

Figure 7.5 Hirsch plot for example

7.5 LIMIT CYCLE COMPUTATIONS

In principle, as mentioned in Section 5.7, it is possible to set up first harmonic balance equations for any configuration of linear and nonlinear elements if one assumes that the inputs to all the nonlinearities are sinusoids of the same frequency but with different phases. For the case of Fig. 7.1 when the bandwidth of the elements of G(s) are low pass and comparable the assumption seems justified. The appropriate DFs for N can be calculated for this situation when N has the general form of eq. (7.1) [25,26]. Alternatively, if initially N and G(s) are both

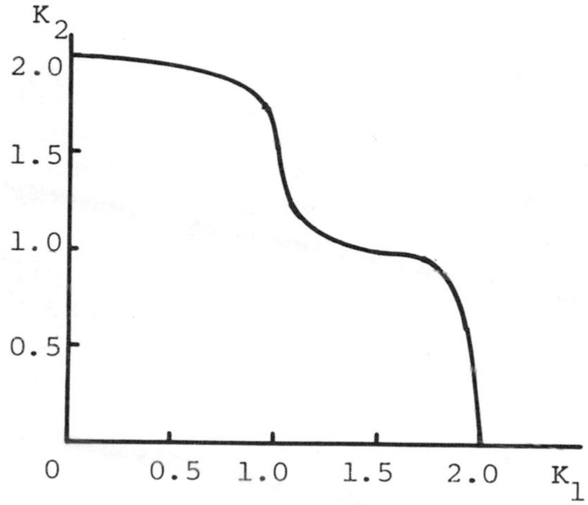

Figure 7.6 Gain plane stability boundary for the example

diagonal and the loops oscillate at different frequencies, the above assumption will not be appropriate for evaluating the system oscillation when small coupling terms are introduced into the off diagonal elements of G(s). Here one could consider combined oscillations of more than one frequency at the nonlinearity inputs but the solution procedure becomes considerably more complex and justification of any DF solution can be difficult.

Assuming a single frequency oscillation the harmonic balance equations for the system of Fig. 7.1 with n=2 and a general multivariable nonlinearity N given by eq. (7.1) are

$$g_{11}M_1 a_1 \cos(\omega t+\psi_1+\theta_{11}) + g_{12}M_2 a_2 \cos(\omega t+\theta+\psi_2+\theta_{12})$$

$$= -a_1 \cos \omega t \tag{7.35}$$

$$g_{21}M_1 a_1 \cos(\omega t+\psi_1+\theta_{21}) + g_{22}M_2 a_2 \cos(\omega t+\theta+\psi_2+\theta_{22})$$

$$= -a_2 \cos(\omega t+\theta) \tag{7.36}$$

where $G_{11}(j\omega) = g_{11}\, e^{j\theta_{11}}$ etc. the nonlinearity inputs are $x_1 = a_1 \cos \omega t$ and $x_2 = a_2 \cos(\omega t+\theta)$ and M_1, M_2, ψ_1 and ψ_2 are the magnitudes and angles of the DFs of N, which for the general case will be functions of a_1, a_2 and θ[26]. When N is diagonal M_1, ψ_1 and M_2, ψ_2 will be functions of only a_1 and a_2, respectively. Eqs. (7.35) and (7.36) provide four nonlinear algebraic equations which can be solved for the unknowns ω, a_1, a_2 and θ. The stability of the solutions can be checked from the stability of the linear multivariable system with the nonlinearities replaced by their IDFs computed for the solution amplitudes.

Reference [27] gives a sequential computational procedure for predicting limit cycles in the system of Fig. 7.1 when N satisfies eq. (7.2). The method uses a harmonic balance approach assuming a single frequency limit cycle and the iterative technique used is assisted by a graphical display of amplitude and frequency dependent polar locus plots. A more general approach for computing limit cycles has been used by Taylor [28] whose algorithm starts from the system equations in nonlinear state variable form. Sine plus bias DF methods are used to compute an assumed single frequency limit cycle. Successful results have been obtained by the procedure for systems with complicated nonlinearities such as multipliers and harmonic functions.

7.6 RELAY SYSTEMS

As indicated in section 6.6 the methods used for investigating limit cycles in chapter 6 can be used for the multivariable system of Fig. 7.1. The nonlinearity N will be assumed to be an uncoupled nonlinearity with all $n_i(x_i)$ of eq. (7.3) relay characteristics. The procedure can however be used when N consists of n^2 individual relay characteristics, that is all $n_{ij}(x_j)$ in eq. (7.2) are relays.

Assuming the nonlinearity outputs are symmetrical odd, one pulse/cycle waveforms of the same fundamental frequency then any limit cycle solution is given by eqs. (6.45) and (6.46), that is

$$\sum_{k=1}^{N} h_k\{A^{\circ}_{G_{jk}}(\alpha_k - \alpha_j, \omega) - A^{\circ}_{G_{jk}}(\gamma_k - \alpha_j, \omega)\} \tag{7.37}$$

has R.P. < 0 and I.P. = $-\pi(\delta_j + \Delta_j)/2$ and

$$\sum_{k=1}^{N} h_k \{A^{\circ}_{G_{jk}}(\gamma_k - \gamma_k, \omega) - A^{\circ}_{G_{jk}}(\alpha_k - \gamma_j, \omega)\} \quad (7.38)$$

has R.P. < 0 and I.P. = $\pi(\delta_j - \Delta_j)/2$. The imaginary parts of these expressions for $j=1, \ldots n$ can be solved for the $2n+1$ unknowns, $\alpha_2 \ldots \alpha_n$, $\gamma_1 \ldots \gamma_n$ and ω assuming $\alpha_1 = 0$.

Recent work [29] has shown that a necessary condition for the stability of any solution is that the linear multivariable system obtained by replacing the relays by their exact incremental gains to the solution inputs must be stable.

Fig. 7.7 shows the output of relay j, assumed to have a dead zone, without and with a small bias signal γ added at the input. The bias output for the second case is

$$(h_j/\pi)\{\Delta\theta_1 + \Delta\theta_2\}$$

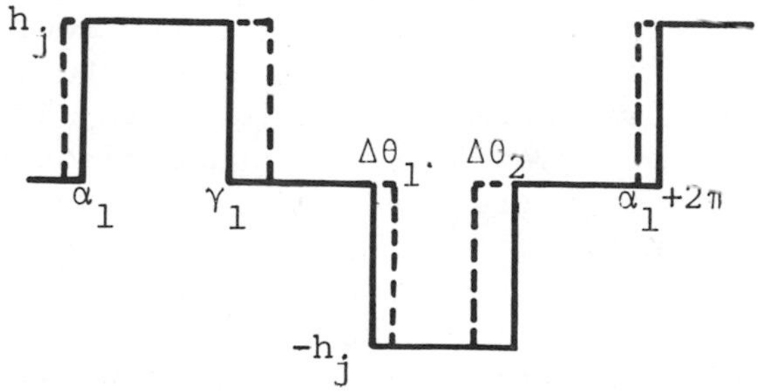

Figure 7.7 Relay output with added bias

But $\Delta\theta_1 = \omega\gamma/\dot{x}(t_j)$ and $\Delta\theta_2 = -\omega\gamma/\dot{x}(t_j + \Delta t_j)$ so that the d.c. incremental gain, N_i, is given by

$$N_i = \frac{h_j \omega}{\pi}\left[\frac{1}{\dot{x}(t_j)} - \frac{1}{\dot{x}(t_j + \Delta t_j)}\right]$$

Since the real parts of the A loci are related to the derivative of x(t) at the switching instants {see eq. (6.14)} it is easily shown that

$$N_i = -\frac{h_i}{2}\left[\frac{1}{R_1} + \frac{1}{R_2}\right]$$

where R_1 and R_2 are the real parts of expressions (7.37) and (7.38) respectively.

To illustrate the above approach we discuss briefly results for a specific example [30]. Consider a system with n=2 containing two ideal relays, that is $\delta_i=\Delta_i=0$ and $h_i=1$ for i=1,2 and a linear transfer function matrix of

$$G(s) = e^{-s}\begin{pmatrix} \frac{1}{s(s+1)} & \frac{\mu}{(s+1)} \\ \frac{\mu}{(s+1)} & \frac{1}{s(s+1)} \end{pmatrix}$$

Since this system is symmetrical as $G_{12}(s) = G_{21}(s)$ and $G_{11}(s) = G_{22}(s)$ one can argue that any oscillations will have equal amplitudes at the relay inputs and the oscillation frequencies can be evaluated from two single loop relay systems with transfer functions $G_{11}(s) \pm G_{12}(s)$, respectively [31]. Use of the describing function method however, gives four not two solutions. These are the two suggested above with the relay input waveforms having equal amplitudes and the output square waves having a phase difference θ of 0° and 180°, respectively, and two where the relay input amplitudes are different and with θ equal to $\pm 90°$. These last two solutions can be shown to occur because Re $G_{12}(j\omega)=0$, for a real ω, which results in an oscillation frequency independent of μ of 0.861 rads/s. Use of the IDF shows that these two solutions are unstable.

The exact analysis using the imaginary part of expression (7.37) yields two nonlinear algebraic equations. Again four solutions are possible with values of θ equal to 0°, 180° and $\pm\ \phi°$, respectively. The phase $\phi°$ is near to 180° but varies with the coupling factor μ. The theoretical results of both the DF method and exact method are summarized in Fig. 7.8. Because of the relatively poor filtering of G(s) the DF frequencies are in error by 5-10% for μ in the range 0 to 1. Simulation results are also included in the figure and are seen to agree with the exact theory. The higher frequency solution, which corresponds to θ=0° is predicted by the incremental gain method to be unstable for $\mu > 0.36$ with the exact solution. The lower frequency solution is particularly interesting since although the frequency error in the DF method is reasonable the solution corresponds to θ=180°, whereas the exact solution is for $\theta=\pm\phi°$. Thus despite the symmetry of the system we have a stable oscillation with the relay inputs unequal in magnitude. The incremental gain method, as expected, predicts the θ=180° solution of the exact analysis to be unstable.

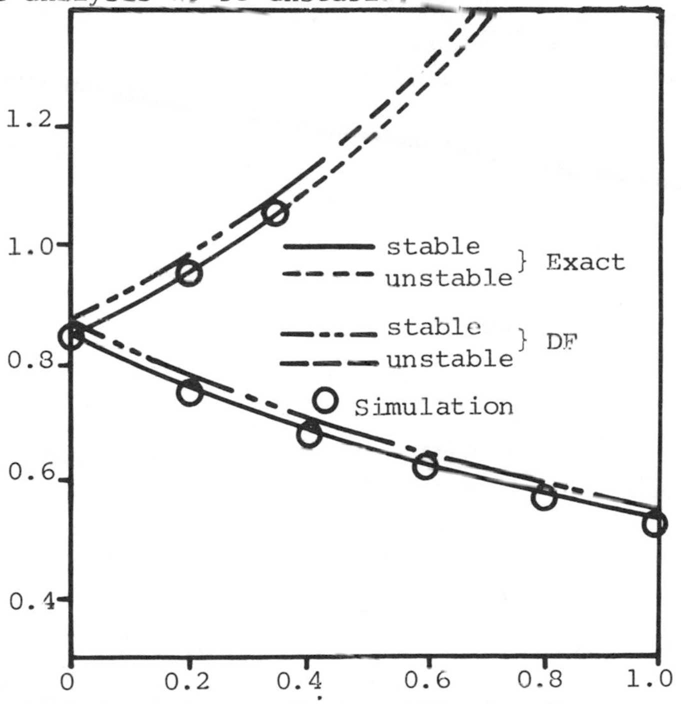

Figure 7.8 Solution for oscillation frequency as a function of the coupling factor

The lower frequency mode is the one attained by the system when released from zero initial conditions. To excite the higher frequency mode the system has to be temporarily forced or given suitable initial conditions. The only difference in the results for μ negative is that the high frequency solution corresponds to a phase θ of 180° not 0°.

7.7 SUMMARY

The aim of this chapter has been to cover available techniques for investigating the stability and determining limit cycles in the nonlinear multivariable system of Fig. 7.1. The coverage of the various methods has necessarily been somewhat brief and for more detailed information the reader is referred to the reference [32], which contains an extensive bibliography. Most of the graphical stability techniques described require diagrams of a complexity which necessitates computer generation but the limit cycle computation methods can be used without an interactive computer terminal. Much work remains to be done to compare the relative merits of the techniques available, "a yardstick of scientific respectability not yet discovered by most control engineering researchers" [33], and to extend the techniques to more complex configurations.

REFERENCES

1. MacFarlane, A.G.J. and Postlethwaite, I., "The generalized Nyquist stability criterion and multivariable root loci", Int. J. Control, 1977, 25, pp. 81-127.

2. Munro, N.(ed),"Modern approaches to control system design", Chapter 4, Peter Peregrinus, London 1979.

3. Tokumaru, H. and Saito, N., "On the absolute stability of automatic control systems with many non-linear characteristics", Memoirs Fac. of Eng., Kyoto Univ., Japan, 1965, 27, pp. 347-379.

4. Partovi, S. and Nahi, E.H., "Absolute stability of dynamic system containing nonlinear functions of several state variables", Automatica, 1969, 5, pp. 465-473.

5. Blight, H.D. and McClamroch, N.H., "Graphical stability criteria for large-scale nonlinear multiloop systems", Proc. 6th IFAC Congress, Boston, 1975, Paper 44.5.

6. Jury, E.I. and Lee, B.W., "The absolute stability of systems with many nonlinearities", Automat. Remote Control, 1965, 26, pp. 913-916.

7. Shankar, S. and Atherton, D.P., "Graphical stability analysis of nonlinear multivariable control systems", Int. J. Control, 1977, 25, pp. 365-388.

8. Mees, A.I., and Atherton, D.P., "The Popov criterion for multiple-loop feedback systems" to be published IEEE Trans. Automat. Contr.

9. Mees, A.I. and Atherton, D.P., "Domains containing the field of values of a matrix", Linear Algebra and Its Applications, 1979, 26, pp. 289-296.

10. Cook, P.A.; "Modified multivariable circle theorems", Recent Mathematical Developments in Control Theory, D.J. Bell, Ed., London, Academic Press, 1973, pp. 367-372.

11. Rosenbrock, H.H., "Multivariable circles theorems", Recent Mathematical Developments in Control Theory, D.J. Bell, Ed., London, Academic Press, 1973, pp. 345-365.

12. Cook, P.A., "Conditions for the absence of limit cycles", IEEE Trans. Automat. Contr., 1976, AC-21, pp. 339-345.

13. Falb, P.L., Freedman, M.I. and Zames, G., "Input-output stability - a general viewpoint", Proc. 4th IFAC Congress, Warsaw, 1969, paper 41.3.

14. Mees, A.I. and Rapp, P.E., "Stability criteria for multiple-loop nonlinear feedback systems", Proc. IFAC Symp. on Multivariable Technological Systems, Fredericton, 1977, pp. 183-188

15. Atherton, D.P. and Owens, D.H., "Boundedness properties of nonlinear multivariable feedback systems", Electronics Letters, 1979, 15, pp. 559-561.

16. El-Sakkary, A.K., "Stability of and oscillations in nonlinear multivariable control systems", M.Sc.E. Thesis, Univ. of New Brunswick, Fredericton, 1978.

17. El Sakkary, A.K. and Atherton, D.P., "Computer graphics methods for nonlinear multivariable systems", IFAC Computer Aided Design Symposium Zurich, 1979, pp. 447-452.

18. Mees, A. "Describing Functions, Circle Criteria and Multiloop Feedback Systems", Proc. IEE, 1973, 120, pp. 126-130.

19. Ramani, N. and Atherton, D.P., "Frequency Response Methods for Nonlinear Multivariable Systems", Proc. Canadian Conference on Automatic Control, Fredericton, 1973, paper 9.2.

20. Ramani, N. and Atherton, D.P., "Stability of non-linear Multivariable Systems", IFAC Symposium on Multivariable Technological Systems, 1974, paper 2-10.

21. Ramani, N., Atherton, D.P., and Mees, A., "Describing Functions, Circle Criteria and Multiloop Feedback Systems", Proc. IEE, 1973, 120, p. 814.

22. Ramani, N., "Some Aspects of the Analysis and Design of Multivariable Systems", Ph.D. Thesis, University of New Brunswick, May 1974.

23. Ramani, N. and Atherton, D.P., "A Describing Function Method for the Approximate Stability of Nonlinear Multivariable Systems", Report SDC 1, E.E. Dept., University of New Brunswick, January 1975.

24. Minc. H. and Marcus, M., "A survey of matrix theory and matrix inequalities", Prindle, Weber & Schmidt Inc., Boston, 1964.

25. Balasubramanian, R. and Atherton, D.P., "Response of multidimensional nonlinearities to inputs which are separable processes", Proc. I.E.E., 1968, 115, pp. 581-90.

26. Atherton, D.P., "Nonlinear Control Engineering: Describing Function Analysis and Design, Van Nostrand Reinhold, London, 1975

27. Gray, J.O. and Al-Janabi, T.H., "The numerical design of multivariable nonlinear feedback systems", Proc. IFAC Symposium on Multivariable Technological Systems, Fredericton, 1977, pp. 233-238.

28. Taylor, J.H., "A new algorithmic limit cycle analysis method for multivariable systems", Report No. TIM-612-2, TASC, Reading Mass., Oct. 1977.

29. Balasubramanian, R., "Stability of Oscillations in Feedback Systems Containing a Relay", to be published in IEE Proceedings-D.

30. Balasubramanian, R. and Atherton, D.P., "Limit Cycles in Systems with Single and Multiple Relays", JACC, San Francisco, August, 1980.

31. Lindgren, A.G. and Pinkos, R.F., "Stability of symmetric nonlinear multivariable systems", J. Franklin Inst., 1966, 282, pp. 92-101.

32. Atherton, D.P. and Dorrah, H.T., "A survey on nonlinear oscillations", Int. J. of Control, 31, pp. 1041.1105, 1980.

33. Atherton, D.P. and Cellier, F., "The value of computer aided design", Report on Round Table Discussion, IFAC Computer Aided Design Symposium, Zurich, August 1979.

PROBLEMS

1. The system of Fig. 7.1 has a diagonal N of two identical relays with dead zone having $\delta=h=1$ and $\Delta=0$. The linear plant has a transfer matrix

$$G(s) = \frac{K}{s(s+1)^2}\begin{pmatrix} 1 & 0.2 \\ 0.3 & 1.5 \end{pmatrix}.$$

Determine the maximum value of K for stability by

(a) The absolute stability method of Mees.
(b) The exact relay method.
(c) The DF method.

2. Repeat problem 1 for

$$G(s) = \frac{K}{s(s+1)^2}\begin{pmatrix} 1 & 0.2 \\ -0.3 & 1.5 \end{pmatrix}.$$

3. Repeat problem 1 for

$$G(s) = \frac{K}{(s+1)^2(s+2)}\begin{pmatrix} 1 & 0.3 \\ 0.4 & 1.5 \end{pmatrix}.$$

4. Repeat problem 1 for

$$G(s) = \frac{K}{(s+1)^2(s+2)}\begin{pmatrix} 1 & -0.3 \\ 0.4 & 1.5 \end{pmatrix}.$$

5. Repeat problems 1 to 4 except for method (b), for N a diagonal nonlinearity with identical saturation characteristics with $m=\delta=1$.

6. Use the DF method to investigate possible limit cycles in a system with a diagonal N having two ideal relays and

$$G(s) = \frac{K}{s(s+1)^2}\begin{pmatrix} 1 & \alpha \\ \alpha & 1 \end{pmatrix}; \quad 0<\alpha<0.7.$$

Are there any solutions for the relays having unequal input sinusoidal amplitudes?

7. Repeat problem 6 for

$$G(s) = \frac{1}{(s+1)^2} \begin{pmatrix} 1/s & \alpha \\ \alpha & 1/s \end{pmatrix}; \quad 0<\alpha<0.7.$$

8. Use the DF method to investigate possible limit cycles in a system with N a diagonal nonlinearity of three ideal relays and

$$G(s) = \frac{1}{(s+1)^3} \begin{pmatrix} 1 & 0.2 & 0.4 \\ 0.4 & 1 & 0.2 \\ 0.2 & 0.4 & 1 \end{pmatrix}.$$

Comments	$N_p(a) = a_1/a$
… nonlinearity $= m_1x$ $x<\delta$ $m_2x+(m_1-m_2)\delta$ $x>\delta$	$N_p = (m_1-m_2)f_1(\delta/a) + m_2$
9. Saturation with dead zone $n(x) = 0$ $x<\delta_1$ $m(x-\delta_1)$ $\delta_1<x<\delta_2$ $m(\delta_2-\delta_1)$ $x>\delta_2$	$N_p = m\{f_1(\delta_2/a) - f_1(\delta_1/a)\}$
10. $n(x) = 0$ $x<\delta_1$ $m_1(x-\delta_1)$ $\delta_1<x<\delta_2$ $m_1(\delta_2-\delta_1)+m_2(x-\delta_2)$ $x>\delta_2$	$N_p = -m_1f_1(\delta_1/a) + (m_2-m_1)f_1(\delta_2/a) + m_2$
11. Limited field of view $n(x) = m_1x$ $x<\delta$ $m_1 - m_2(x-\delta)$ $\delta<x<(m_1+m_2)\delta/m_2$ 0 $x>(m_1+m_2)\delta/m_2$	$N_p = (m_1+m_2)f_1(\delta/a) - m_2f_1\{(m_1+m_2)\delta/m_2a\}$

Appendix I

Tables of DFs and SBDFs

Short tables of DFs and SBDFs are given in this appendix. For more extensive tables the reader is referred to references [1,2]. All the nonlinearities in both tables are assumed to be odd symmetrical and the mathematical definitions given are for $x>0$.

For Table I.1 the input $x = a\cos\theta$ and the following relationships apply.

$$a_1 = (2/\pi)\int_0^{\pi} y(\theta)\cos\theta \, d\theta = (4/a)\int_0^{a} x\, n_p(x)p(x)\, dx$$

$$b_1 = (2/\pi)\int_0^{\pi} y(\theta)\sin\theta \, d\theta = (4/a)\int_0^{a} n_q(x)\, dx$$

where $p(x) = (1/\pi)(a^2-x^2)^{-1/2}$

The DF, $N(a)$, is given by

$$N(a) = (a_1 - jb_1)/a = N_p(a) + jN_q(a).$$

Use is made in the Table of the saturation function

$$f_1(\rho) = \begin{cases} 1 & \text{for } \rho > 1 \\ (2/\pi)\{\sin^{-1}\rho + \rho(1-\rho^2)^{1/2} & \text{for } |\rho| < 1 \\ -1 & \text{for } \rho < -1 \end{cases}$$

For Table I.2 the input is $\gamma + a\cos\theta$ and the SBDFs for SVNLs are given by

$$N_{\gamma}(a,\gamma) = (1/\gamma)\int_{-a}^{a} n(x+\gamma)p(x)\, dx$$

$N(a,\gamma) = (2/a^2)\int_{-a}^{a} x\, n(x+\gamma)p(x)\, dx$

In addition to $f_1(\rho)$ defined above the following functions are also used in Table I.2.

$$g(\rho) = \begin{cases} -1/2 & \rho < -1 \\ (1/\pi)\sin^{-1}\rho & |\rho| \le 1 \\ 1/2 & \rho > 1 \end{cases}$$

$$D_1(\rho) = \begin{cases} (1-\rho^2)^{1/2} & |\rho| \le 1 \\ 0 & |\rho| > 1 \end{cases}$$

$$f_0(\rho) = \begin{cases} (2/\pi)\{\rho \sin^{-1}\rho + (1-\rho^2)^{1/2}\} & |\rho| \le 1 \\ |\rho| & |\rho| > 1 \end{cases}$$

The numbers used in this table refer to the nonlinearities with the same numbers as Table I.1.

TABLE I.1

Table of DFs

(a) SVNLs

Nonlinearity	Comments	$N_p(a) = a_1/a$
1. General quantiser See Section 4.2.2	$a < \delta_1$	$N_p = 0$
	$\delta_{M+1} > a > \delta_M$	$N_p = (4/a^2\pi)\sum_{m=1}^{M} h_m(a^2-\delta_m^2)^{1/2}$
2. …iform quantiser h… = …h …δ/2	$a < \delta$	$N_p = 0$
	$(2M+1)\delta > a > (2M-1)\delta$	$N_p = (4h/a^2\pi)\sum_{m=1}^{M}(a^2-n^2\delta^2)^{1/2}$
…ne	$a < \delta$	$N_p = 0$
	$a > \delta$	$N_p = 4h(a^2-\delta^2)^{1/2}/a^2\pi$
		$N_p = 4h/a\pi$
		$N_p = (4h/a\pi) + m$
		$N_p = mf_1(\delta/a)$
		…$\{1-f_1(\delta/a)\}$

SVNLs (continued)

Nonlinearity
8. Gain changin… n(x)…

SVNLs (continued)

Nonlinearity	Comments	$N_p(a)=a_1/a$
12. $n(x)=x^m$	$m>-2$ Γ is the gamma function	$N_p = \frac{\Gamma(m+1)\,a^{m-1}}{2^{m-1}\Gamma[(3+m)/2]\Gamma[(1+m)/2]}$ $= \frac{2}{\sqrt{\pi}} \frac{\Gamma[(m+2)/2]a^{m-1}}{\Gamma[(m+3/2]}$
13. $n(x)=x^n$	n odd integer, μ_n, is the nth moment of the amplitude distribution function of a sinusoid	$N_p = \frac{n(n-2)(n-4)\ldots 3}{(n+1)(n-1)\ldots 4}\,a^{n-1}$ $= \frac{2}{a^2}\,\mu_{n+1}$
14. $n(x) = x^n$	n even integer	$N_p = \frac{4}{\pi}\,\frac{n(n-2)\ldots 2}{(n+1)(n-1)\ldots 3}\,a^{n-1}$
15. Harmonic nonlinearity $n(x)=A_m \sin mx$	$J_s(ma)$ is the Bessel function of order s	$N_p = 2A_m J_1(ma)/a$
16. $n(x) = A_m \sinh mx$	$I_s(ma)$ is the modified Bessel function of order s	$N_p = 2A_m I_1(ma)/a$

(a) SVNLs (continued)

Nonlinearity	Comments	$N_p(a) = a_1/a$
17. Exponential saturation $n(x) = 1-e^{-cx}$	$I_1(ca)$ is the modified Bessel function of order 1. $S_1(ca)$ is the modified Struve function of order 1.	$N_p = (2/a)\{I_1(ca)-S_1(ca)\}$
18. $n(x) = \dfrac{cx}{(1+c^2x^2)}$	K(k) and E(k) are complete elliptic integrals $u = ca/(1+c^2a^2)^{1/2}$	$N_p = (4/\pi)\{(-u/c^2a^3)K(u) + (1/au)E(u)\}$

(b) DVNLs

Nonlinearity	Comments	$N_p(a) = a_1/a$	$N_q(a) = -b_1/a$
19. On-off relay with hysteresis Fig. 4.3 with $\delta = 0$	$a < \Delta$	$N_p = 0$	$N_q = 0$
	$a > \Delta$	$N_p = 4h(a^2-\Delta^2)^{1/2}/a^2\pi$	$N_q = -4h\Delta/a^2\pi$
20. On-off relay with dead zone and hysteresis Fig. 4.3	$a < \Delta +$	$N_p = 0$	$N_q = 0$
	$a > \Delta +$	$N_p = \dfrac{2h}{a^2\pi}[\{a^2-(\delta+\Delta)^2\}^{1/2} + \{a^2-(\delta-\Delta)^2\}^{1/2}]$	$N_q = -4h\Delta/a^2\pi$

TABLE I.2

Nonlinearity	$N_\gamma(a,\gamma)$ and $N(a,\gamma)$
2	$N_\gamma=(h/\gamma)\sum_{m=1}^{M}\{g(\frac{n\delta+\gamma}{a}) - g(\frac{n\delta-\gamma}{a})\}$ $N_p=(2h/a\pi)\sum_{m=1}^{M}\{D_1(\frac{n\delta+\gamma}{a}) + D_1(\frac{n\delta-\gamma}{a})\}$
3	$N_\gamma=(h/\gamma)\{g(\frac{\delta+\gamma}{a}) - g(\frac{\delta-\gamma}{a})\}$ $N_p=(2h/a\pi)\{D_1(\frac{\delta+\gamma}{a}) + D_1(\frac{\delta-\gamma}{a})\}$
4	$N_\gamma=(2h/\gamma)g(\gamma/a)$ $N_p=(4h/a\pi)D_1(\gamma/a)$
5	$N_\gamma=m + (2h/\gamma)g(\gamma/a)$ $N_p=m + (4h/a\pi)D_1(\gamma/a)$
6	$N_\gamma=(ma/2\gamma)\{f_0(\frac{\delta+\gamma}{a}) - f_0(\frac{\delta-\gamma}{a})\}$ $N_p=(m/2)\{f_1(\frac{\delta+\gamma}{a}) + f_1(\frac{\delta-\gamma}{a})\}$
7	$N_\gamma=m - (ma/2\gamma)\{f_0(\frac{\delta+\gamma}{a}) - f_0(\frac{\delta-\gamma}{a})\}$ $N_p=m - (m/2)\{f_1(\frac{\delta+\gamma}{a}) + f_1(\frac{\delta-\gamma}{a})\}$
8	$N_\gamma=\{(m_1-m_2)a/2\gamma\}\{f_0(\frac{\delta+\gamma}{a}) - f_0(\frac{\delta-\gamma}{a})\} + m_2$ $N_p=\{(m_1-m_2)/2\}\{f_1(\frac{\delta+\gamma}{a}) + f_1(\frac{\delta-\gamma}{a})\} + m_2$

Nonlinearity	$N_\gamma(a,\gamma)$ and $N(a,\gamma)$	
13	$N_\gamma = \sum_{k=0}^{n} {}_nC_k \mu_k \gamma^{n-k-1}$	${}_nC_k$ is the binomial coefficient
	$N_p = (2/a^2) \sum_{k=0}^{n} {}_nC_k \mu_{k+1} \gamma^{n-k}$	
15	$N_\gamma = (1/\gamma) A_m J_0(ma)$ sin $m\gamma$	
	$N_p = (2/a) A_m J_1(ma)$ cos $m\gamma$	
16	$N_\gamma = (1/\gamma) A_m I_0(ma)$ sinh $m\gamma$	
	$N_p = (2/a) A_m I_1(ma)$ cosh $m\gamma$	

REFERENCES

1. Atherton, D.P.: "Nonlinear control engineering", Van Nostrand Reinhold, London, 1975, appendix A1.

2. Gelb, A. and Vander Velde, W.E.: "Multiple-input describing functions and nonlinear system design", McGraw-Hill, New York, 1968, appendix B.

Appendix II
Tables of A Loci

The A locus of a transfer function, G(s), is defined by

$$A_G(\theta,\omega) = \mathrm{Re}\, A_G(\theta,\omega) + j\, \mathrm{Im}\, A_G(\theta,\omega)$$

with

$$\mathrm{Re}\, A_G(\theta,\omega) = \sum_{n=1}^{\infty} V_G(n\omega)\sin n\theta + U_G(n\omega)\cos n\theta$$

$$\mathrm{Im}\, A_G(\theta,\omega) = \sum_{n=1}^{\infty} (1/n)\{V_G(n\omega)\cos n\theta - U_G(n\omega) \sin n\theta\}.$$

When the locus is obtained by summing over odd values of n only it is denoted by A°. The following tables show how A loci can be evaluated for various transfer functions.

TABLE II.1

A° Loci

Type No.	Re $A^\circ_G[\theta,\omega]$	Im $A^\circ_G[\theta,\omega]$
1	$-\lambda^{-1}S^\circ_{-1,0}$	$-\lambda^{-1}C^\circ_{-2,0}$
2	$C^\circ_{0,1}-\lambda S_{1,1}$	$-S^\circ_{-1,1}-\lambda C^\circ_{0,1}$
3	$-\lambda^{-1}S^\circ_{-1,1}-C^\circ_{0,1}$	$-\lambda^{-1}C^\circ_{-2,0}+S^\circ_{-1,1}+C^\circ_{0,1}$
4	$-\lambda^{-2}C^\circ_{-2,0}$	$\lambda^{-2}S^\circ_{-3,0}$
5	$-2\lambda S^\circ_{1,2}+C^\circ_{0,2}-\lambda^2C^\circ_{2,2}$	$-2\lambda C^\circ_{0,2}-S^\circ_{-1,2}+\lambda^2S^\circ_{1,2}$
6	$C^\circ_{0,1}-\lambda S^\circ_{1,1}+2\lambda S^\circ_{1,2}$ $-C^\circ_{0,2}-\lambda^2C^\circ_{2,2}$	$-S^\circ_{-1,1}-\lambda C^\circ_{0,1}+2\lambda C^\circ_{0,2}$ $+S^\circ_{-1,2}-\lambda^2S^\circ_{1,2}$
7	$\lambda^{-3}S^\circ_{-3,0}$	$\lambda^{-3}C^\circ_{-4,0}$
8	$C^\circ_{0,2}-4\lambda^2C^\circ_{2,3}-4\lambda S^\circ_{1,3}$ $+\lambda S^\circ_{1,2}$	$-3\lambda C^\circ_{0,2}+4\lambda^3C^\circ_{2,3}-S^\circ_{-1,2}$ $+4\lambda^2S^\circ_{1,3}$
9	$\frac{\pi}{4b\omega}\{\frac{c_1}{d}\}$	$\frac{\pi}{4\omega_0^2}\left[\frac{b(c_2-d)-ac_1}{bd}\right]$
10	$-\frac{\pi}{4b\omega}\left[\frac{ac_1+bc_2}{d}\right]$	$\frac{\pi}{4b}\{\frac{c_1}{d}\}$

Notation

The sine and cosine series $S^\circ_{j,k}$ and $C^\circ_{j,k}$ are defined as follows:

$$S^{\circ}_{j,k} = S^{\circ}_{j,k}(\theta,\lambda) = \sum_{n=1(2)}^{\infty} \frac{n^j \sin n\theta}{(1+n^2\lambda^2)^k} \qquad \text{for } 0 < \theta < \pi$$

$$S^{\circ}_{j,k}(-\theta,\lambda) = -S^{\circ}_{j,k}(\theta,\lambda)$$

$$S^{\circ}_{-1,0}(\theta,\lambda) = \pi/4$$

$$S^{\circ}_{1,1}(\theta,\lambda) = \pi\cosh[\pi-2\theta)/2\lambda]/4\lambda^2 \cosh(\pi/2\lambda)$$

$$S^{\circ}_{-1,1}(\theta,\lambda) = S^{\circ}_{-1,0} - \lambda^2 S^{\circ}_{1,1}$$

$$\begin{aligned} S^{\circ}_{1,2}(\theta,\lambda) = {} & \{\pi\theta \sinh[(\pi-\theta)/2\lambda] \\ & + \pi^2 \sinh(\theta/\lambda)/2 \cosh(\pi/2\lambda)\}/8\lambda^3 \\ & \times \cosh(\pi/2\lambda) \end{aligned}$$

$$S^{\circ}_{-1,2}(\theta,\lambda) = S^{\circ}_{-1,1} - \lambda^2 S^{\circ}_{1,2}$$

$$\begin{aligned} S^{\circ}_{1,3}(\theta,\lambda) = {} & \{\pi^2\lambda \sinh(\theta/\lambda) + 2\pi\theta\lambda \cosh(\pi/2\lambda) \\ & \times \sinh[(\pi-2\theta)/2\lambda] - 2\pi^2\theta \cosh(\theta/\lambda) \\ & + 2\pi\theta^2 \cosh(\pi/2\lambda)\cosh[(\pi-2\theta)/2\lambda] \\ & + \pi^3 \sinh(\theta/\lambda)\tanh(\pi/2\lambda)\}/64\lambda^4 \cosh^2(\pi/2\lambda) \end{aligned}$$

$$S^{\circ}_{-3,0}(\theta,\lambda) = (\pi^2\theta - \pi\theta^2)/8$$

and

$$C^{\circ}_{j,k} = C^{\circ}_{j,k}(\theta,\lambda) = \sum_{n=1(2)}^{\infty} \frac{n^j \cos n\theta}{(1+n^2\lambda^2)^k} \qquad \text{for } 0 < \theta < \pi$$

$$C^{\circ}_{j,k}(-\theta,\lambda) = C^{\circ}_{j,k}(\theta,\lambda)$$

$$C^{\circ}_{0,1}(\theta,\lambda) = \{\pi \sinh[(\pi-2\theta)/2\lambda]\}/4\lambda \cosh(\pi/2\lambda)$$

$$C^{\circ}_{-2,0}(\theta,\lambda) = (\pi^2-2\pi\theta)/8$$

$$C^{\circ}_{2,2}(\theta,\lambda) = \{\pi\lambda \sinh[(\pi-2\theta)/2\lambda] - \pi\theta \cosh[(\pi-2\theta)/2\lambda] + [\pi^2 \cosh(\theta/\lambda)]/2\cosh(\pi/2\lambda)\}/8\lambda^4\cosh(\pi/2)$$

$$C^{\circ}_{0,2}(\theta,\lambda) = C^{\circ}_{0,1} - \lambda^2 C^{\circ}_{2,2}$$

$$C^{\circ}_{2,3}(\theta,\lambda) = \{2\pi\lambda^2\sinh[(\pi-2\theta)/2\lambda)]\cosh(\pi/2\lambda) - \pi^2\lambda \cosh(\theta/\lambda) + 2\pi\theta\lambda \cosh[(\pi-2\theta)/2\lambda] \cosh(\pi/2\lambda) - 2\pi^2\theta \sinh(\theta/\lambda) - 2\pi\theta^2\sinh[(\pi-2\theta)/2\lambda]\cosh(\pi/2\lambda) + \pi^3 \cosh(\theta/\lambda)\tanh(\pi/2\lambda)\}/64\lambda^5 \cosh^2(\pi/2\lambda)$$

$$C^{\circ}_{-4,0}(\theta,\lambda) = (\pi^4 - 6\pi^2\theta^2 + 4\pi\theta^3)/96$$

$$C^{\circ}_{2,1}(\theta,\lambda) = \{-\pi \sinh[(\pi-2\theta)/2\lambda]\}/4\lambda^3 \cosh(\pi/2\lambda).$$

Also, for the transfer functions with complex poles

$$a = \zeta\omega_o, \quad b = \omega_o\sqrt{1 - \zeta^2}$$

$$d = \cosh(a\pi/\omega) + \cos(b\pi/\omega)$$

$$c_1 = \exp(a\theta/\omega)\{\exp(-a\pi/\omega)\sin(b\theta/\omega) - \sin[(\pi-\theta)b/\omega]\}$$

$$c_2 = \exp(a\theta/\pi)\{\exp(-a\pi/\omega)\cos(b\theta/\omega) + \cos[(\pi-\theta)b/\omega]\}$$

TABLE II.2

A Loci

Type	Re $A_G[\theta,\omega]$	Im $A_G[\theta,\omega]$
1	$-\lambda^{-1}S_{-1,0}$	$-\lambda^{-1}C_{-2,0}$
2	$C_{0,1}-\lambda S_{1,1}$	$-S_{-1,1}-\lambda C_{0,1}$
3	$-\lambda^{-1}S_{-1,1}-C_{0,1}$	$-\lambda^{-1}C_{-2,0}+S_{-1,1}+\lambda C_{0,1}$
4	$-\lambda^{-2}C_{-2,0}$	$\lambda^{-2}S_{-3,0}$
5	$-2\lambda S_{1,2}+C_{0,2}-\lambda^{2}C_{2,2}$	$-2\lambda C_{0,2}-S_{-1,2}+\lambda^{2}S_{1,2}$
6	$C_{0,1}-\lambda S_{1,1}+2\lambda S_{1,2}$ $-C_{0,2}-\lambda^{2}C_{2,2}$	$-S_{-1,1}-\lambda C_{0,1}+2\lambda C_{0,2}$ $+S_{-1,2}-\lambda^{2}S_{1,2}$
7	$\lambda^{-3}S_{-3,0}$	$\lambda^{-3}C_{-4,0}$
8	$C_{0,2}-4\lambda^{2}C_{2,3}-4\lambda S_{1,3}$ $+\lambda S_{1,2}$	$-3\lambda C_{0,2}+4\lambda^{3}C_{2,3}-S_{-1,2}$ $+4\lambda^{2}S_{1,3}$
9	$-\left(\frac{1}{2\omega_0^2}+\frac{\pi c_3 d_1}{2b\omega d_0}\right)$	$\frac{1}{\omega_0^2}\left(-(\mathrm{sgn}\,\theta)\frac{\pi-\lvert\theta\rvert}{2}+\frac{\omega a}{\omega_0^2}+\frac{\pi[c_1-c_2]d_1}{2bd_0}\right)$
10	$\frac{\pi[c_4-c_5]d_1}{2b\omega d_0}$	$-\left(\frac{\omega}{2\omega_0^2}+\frac{\pi c_3 d_1}{2bd_0}\right)$

The sine and cosine series $S_{j,k}$ and $C_{j,k}$ are defined as follows:

$$S_{j,k} = S_{j,k}(\theta,\lambda) = \sum_{n=1}^{\infty} \frac{n^j \sin n\theta}{(1+n^2\lambda^2)^k} \qquad \text{for } 0 < \theta < 2\pi$$

$$S_{j,k}(-\theta,\lambda) = -S_{j,k}(\theta,\lambda)$$

$$S_{-1,0}(\theta,\lambda) = (\pi-\theta)/2$$

$$S_{1,1}(\theta,\lambda) = (\pi/2\lambda)^2 \sinh[(\pi-\theta)/\lambda]/\sinh(\pi/\lambda)$$

$$S_{-1,1}(\theta,\lambda) = S_{-1,0}(\theta,\lambda) - \lambda^2 S_{1,1}(\theta,\lambda)$$

$$S_{1,2}(\theta,\lambda) = -(\pi/4\lambda^3)(\pi \sinh(\theta/\lambda) - \theta \sinh(\pi/\lambda)\cosh \times[(\pi-\theta)/\lambda])/\sinh^2(\pi/\lambda)$$

$$S_{-1,2}(\theta,\lambda) = S_{-1,1}(\theta,\lambda) - \lambda^2 S_{1,2}(\theta,\lambda)$$

$$\begin{aligned} S_{1,3}(\theta,\lambda) = -(\pi/16\lambda^4)((&\theta \sinh(\pi/\lambda)\cosh[(\pi-\theta)/\lambda] \\ &- \pi \sinh(\theta/\lambda))(\lambda \sinh(\pi/\lambda) + 2\pi \cosh(\pi/\lambda)) \\ &+ \sinh(\pi/\lambda)(\pi\theta \cosh(\theta/\lambda) \\ &- \pi\theta \cosh[(2\pi-\theta)/\lambda] + \theta^2 \sinh(\pi/\lambda) \\ &\times \sinh(\pi-\theta)/\lambda]))/\sinh^3(\pi/\lambda) \end{aligned}$$

$$S_{-3,0}(\theta,\lambda) = [(\theta-\pi)^3 - \pi^2\theta + \pi^3]/12$$

and

$$C_{j,k} = C^{\circ}_{j,k}(\theta,\lambda) = \sum_{n=1}^{\infty} \frac{n^j \cos n\theta}{(1+n^2\lambda^2)^k} \qquad \text{for } 0 \le \theta < 2\pi$$

$$C_{j,k}(-\theta,\lambda) = C_{j,k}(\theta,\lambda)$$

$$C_{0,1}(\theta,\lambda) = (\pi/2\lambda)\cosh[(\pi-\theta)/\lambda]/\sinh(\pi/\lambda) - (1/2)$$

$$C_{-2,0}(\theta,\lambda) = (\theta-\pi)^2/4 - \pi^2/12$$

$$C_{2,2}(\theta,\lambda) = -(\pi/4\lambda^4)(\pi \cosh(\theta/\lambda) - \{\lambda \sinh(\pi/\lambda)\cosh[(\pi-\theta)/\lambda] - \theta \sinh(\pi/\lambda)\sinh[(\pi-\theta)/\lambda]\}) \div \sinh^2(\pi/\lambda)$$

$$C_{0,2}(\theta,\lambda) = C_{0,1}(\theta,\lambda) - \lambda^2 C_{2,2}(\theta,\lambda)$$

$$C_{2,3}(\theta,\lambda) = \frac{\partial}{\partial\theta}[S_{1,3}(\theta,\lambda)]$$

$$C_{-4,0}(\theta,\lambda) = \{2\pi^2(\theta-\pi)^2 - (\theta-\pi)^4\}/48 - (7\pi^4/720)$$

Also, for the transfer functions with complex poles

$$a = \zeta\omega_o, \quad b = \omega_o\sqrt{1-\zeta^2}$$

$$d_0 = \cosh 2\pi a/\omega - \cos 2\pi b/\omega$$

$$d_1 = \exp[(\mathrm{sgn}\theta)a|\theta|/\omega]$$

$$c_1 = \{a \sin[b(|\theta| - 2\pi)/\omega] - (\mathrm{sgn}\theta)b \cos[b(|\theta| - 2\pi)/\omega]\}$$

$$c_2 = \{a \sin b|\theta|/\omega - (\mathrm{sgn}\theta)b \cos b|\theta|/\omega\}\{\exp[-(\mathrm{sgn}\theta)2\pi a/\omega]\}$$

$$c_3 = (\sin[b(|\theta| - 2\pi)/\omega] - \sin b|\theta|/\omega\{\exp[-(\mathrm{sgn}\theta)2\pi a/\omega]\})$$

$$c_4 = \{a \sin[b(|\theta| - 2\pi/\omega] + (\mathrm{sgn}\theta)b \cos[b(|\theta| - 2\pi/\omega]\}$$

$$c_5 = \{a \sin b|\theta|/\omega + (\mathrm{sgn}\theta)b \cos b|\theta|/\omega\}\{\exp[-(\mathrm{sgn}\theta)2a/\omega]\}$$

Index